Biotechnology In the Time of COVID-19

Commentaries from the Front Line

Edited by
Dr. Jeremy M. Levin

One year ago
Jerry
2021

RosettaBooks®
NEW YORK 2020

Jeremy M. Levin

Stéphane Bancel

Sol Barer

Jeff Berkowitz

Karen Bernstein

Bruce Booth

Brook Byers

Ron Cohen

Samantha Du

Deborah Dunsire

Wainwright Fishburn

Cedric François

Frederick Frank

Julie Louise Gerberding

James Greenwood

Paul Hastings

Quita Highsmith

Steven Holtzman

Alexander Karnal

Jeff Kindler

Rachel King

Nina Kjellson

Peter Kolchinsky

Mark Lampert

John Maraganore

Kiran Mazumdar-Shaw

Michelle McMurry-Heath

Philip Miller

Kenneth Moch

John V. Oyler

Stelios Papadopoulos

Andy Plump

Richard Pops

Geoff Porges

Luke Rosen

Vicki L. Sato

George Scangos

Paul Sekhri

Christi Shaw

Bill Sibold

Jeffrey M. Solomon

Jean-Pierre Sommadossi

Otello Stampacchia

Samuel Waksal

Yaron Werber

Sylvia Wulf

John Young

Alex Zhavoronkov

Biotechnology in the Time of COVID-19

First edition published 2020 by RosettaBooks

Cover design by Lon Kirschner

www.kirschnerdesign.com

ISBN-13 (print): 978-0-7953-5297-3
ISBN-13 (ebook): 978-0-7953-5298-0

Library of Congress Control Number: 2020939728

RosettaBooks®

www.RosettaBooks.com

Printed in the United States of America

CONTENTS

The View from Inside Companies

Broad Lessons

Leadership

Investor Perspectives

Looking to the Future

FROM THE EDITOR

Each of the contributors to this book was gracious enough to participate without compensation. I have assigned my royalties equally to two nonprofits dedicated to research into 7,000 rare diseases: globalgenes.org and lifechangingmedicines.org.

RosettaBooks will join me in donating a percentage of their revenues equally to the same nonprofits.

About Global Genes: Our story began with our loved ones. We are the friends, family, and supporters of patients close to us who are affected by rare disease. Rare diseases afflict more than 30 million Americans and impact over 400 million people worldwide. Spanning over 8,000 diseases, this community is the largest patient population on the planet. That's the reason Global Genes was born—to connect, empower, and inspire the rare disease patient advocates who decided they were going to forge a new reality for themselves and loved ones, to not take that line as a *fait accompli*. From educating the newly diagnosed patients and caregivers to building technical capacity across our 600-plus patient advocacy organizations, Global Genes champions patient-centric approaches to shorten the diagnostic odyssey, increase the number of rare diseases researched, and improve access to treatments for all. Global Genes programs, partnerships, and collaborations foster a robust, scaled R&D landscape for the rare disease community. Together we are allies in rare disease.

About The Institute for Life Changing Medicines: Our story is part of the larger American biotech story—that we are connected as one fighting against disease across the world. As an industry, we face the stark realities of cost, risk, and a need for financial viability. The result is many less common, yet lethal, diseases have not been defeated. Now that problem is addressed by The Institute for Life Changing Medicines. A nonprofit founded by Alex Karnal, Dr. James Wilson, and Dr. Tachi Yamada, which promises to transform the status quo. Together, they employ the collective advantages of nonprofit capital, industry partnerships, and a self-sustaining model to act on their belief that a healthy life is a basic human right. With their efforts, the reach of our industry's mission will be expanded to include the diseases currently left behind. Together, locked arm-in-arm with its for-profit colleagues and partners, we will be even more connected as one industry with one goal, to transform the lives of anyone suffering anywhere in the world.

ACKNOWLEDGMENTS

At the risk of further delay, I ask your forbearance as I thank those who are responsible for conceiving this book and urging me to take this on. I owe them the profound thanks they are due!

Foremost amongst these is Arthur Klebanoff, Chief Executive of RosettaBooks, who a mere four weeks ago casually conceptualized this project, prodded me to take it on, and then pulled off publication on this uber-accelerated schedule.

Similarly, this book could not have occurred without the extraordinary, immediate response of the authors, an exceptional group of leaders who in the midst of their day jobs driving the engine of biotechnology took time out to write these amazing insights. They are representative of the many I was unable to include, whom I also admire deeply and whom each has his or her own story to tell of this period. To those in this book whose sustained friendships stretch over decades, I value you deeply. You make rich the fabric of my life in biotechnology.

My work in medicine and biotechnology would not have been possible without the inspiration I derived from lifelong friends Yoram Hessel, Yehiam Mart, and Eliezer Kalina. Thanks also to Sir Stuart Hampshire who opened the door for me to explore the wonderful world of science despite my then evident lack of academic accomplishment!

Thank you to my family for their belief and their tolerance, and thank you to my mother, Leah, for instilling in me the values that drive me. She is today alone in London, like many others, braving the pandemic. She and my father Archie modeled the value that one never gives up when battling for what is right.

On behalf of all the writers, I offer our deepest condolences to the families worldwide who have lost loved ones to COVID-19.

In addition, as we launch this book against the backdrop of the tragedy of this past explosive and very sad week in America, we remind ourselves of the positive impact we as individuals and as an industry can have not just against disease but more broadly by exemplifying and expressing the standards and values we believe in.

INTRODUCTION: A FORESEEABLE CRISIS, AN UNNECESSARY CATASTROPHE, AND A REMARKABLE RESPONSE

Jeremy M. Levin

History will record this period as one of deep pain and suffering coupled with uncertainty and fear. But it will also record the heroes of today: the first responders, the nurses, the doctors, and those who care for their loved ones who are sick and at home. They faced a monster that swept across borders with impunity. They were alone initially.

Before this happened, and while the COVID-19 crisis was unfolding in China, in January 2020 I traveled to the Far East for the trip of a lifetime. As I entered Cambodia to visit the sites that I had dreamed of, I was struck by the emptiness of the streets. There were few tourists. The glorious ruins of Siem Reap were deserted. The fabled Ta Prohm Temple—inspiration for the mythical "Temple of Doom"—was left unattended and enclosed only by the roots of giant trees. There was fear in the eyes of everyone in the streets.

As I walked the streets and ate in the deserted restaurants, it was clear that this part of the world knew a storm was coming in the form of a dreaded pandemic. I was certain America needed to awaken to this growing crisis, seemingly far from our shores and national consciousness, but clearly heading our way.

I returned to the United States and immediately contacted Jim Greenwood, the CEO of the Biotechnology Innovation Organization (BIO). I was in my eighth month of as chair of BIO and was anxious to know what steps BIO was taking to organize an industry-wide response to COVID-19, though it was still thought to be far offshore and not yet a threat to the world. I asked Jim to determine what our colleagues were doing. Hearing that there was no industry-wide coordination, I recruited Dr. George Scangos of Vir Biotechnology, a deeply experienced scientist and leader, to chair BIO's COVID-19 efforts. Jim, George, and BIO galvanized an industry-wide effort whose story will be told in this book.

This book is the opening chapter of a story of how the biopharmaceutical industry, both large and small companies, rose to battle the COVID-19 virus. It is told by the leaders of that industry in their own words. The industry has been built over more than forty years, an ecosystem of scientists, entrepreneurs, doctors, investors, venture capitalists, analysts, media, and bankers, all committed to advancing global health. In two months, leaders and companies from this ecosystem pivoted from the work they were carrying on pre-COVID-19 to face the virus.

We are in a battle that pits viral biology against human intellect and capability. It is a battle that has been waged before. In the Middle Ages, the Black Death brought Europe to its knees. In the sixteenth and seventeenth centuries, smallpox devastated the native inhabitants of North and South America. In the twentieth century, the Spanish flu swept the globe, killing millions. And in the twenty-first century, SARS, H1N1, and Ebola reared their heads. COVID-19 was a crisis anticipated by many, but one for which we were inadequately prepared. That has led to an unnecessary catastrophe.

Against this backdrop, it is heartrending to see the devasting and capricious effect of this virus as it kills and injures its victims, upends economies, destabilizes social discourse, and engenders political polemic, enabling those with questionable motives to push aside science in the interest of politics. The result is global dislocation and a break in the pattern of all our lives.

Now is the time to shine a light on those who bring hope that we will push back the virus with science and medicine. My optimism that we will succeed is born of the conviction that the biotechnology industry has been indirectly preparing for this moment for years. And now it has stepped forward voluntarily in a massive effort to confront and turn back the virus based on rigorous scientific and medical practice.

This book is a collection of the voices from the front of that fight. It is a compilation of the personal observations of forty-seven leaders across the world of biotechnology. Some describe how the virus hit them personally, or how it impacts their work daily. Others share their observations of the mobilization and preparedness of their companies and the industry. Still others look to the lessons we have learned and to the future. All share the commitment to fight and the conviction that science and medicine will prevail.

As for me, I head a company not directly involved with antiviral treatments, but one that exemplifies the scientific creativity and dedication of the industry. In the face of the pandemic, we know our work must continue. Ovid Therapeutics is a small company focused on transforming the lives of those with rare diseases of the brain. We work with those who are born with and suffer from lifelong conditions that in some cases affect the natural development of the brain and in others manifest themselves in devastating epilepsies that progressively destroy the brain.

Like many biotech firms, our ongoing clinical trials were complicated by COVID-19, and, like many, we had to adjust in an instant to an environment in which patients could not get to clinical trial sites or those sites were closed. Ovid employees, all working from home by early March 2020, rose to the occasion, concerned but undeterred by the virus. By March 20, 2020, as the pandemic raged and we went into lockdown, we announced positive data in our Phase 2 trial in the very rare epilepsies, CDKL5 and Dup15Q. Then, on May 7, 2020, still locked down, we announced positive data in our Phase 2 trial in Fragile X. Looking to the future, we plan to deliver results from a large Phase 2 global trial in the rare epilepsies Dravet and Lenox Gastaut syndromes in the next few months, all on time. And finally, later in the year we expect to deliver results from our pivotal Phase 3 trial in Angelman syndrome.

We found a way to bolster the hope entrusted in us by those that depend on our finding new medicines. Undaunted, many other biotechnology companies in the middle of trials, like Ovid, continue with determination to seek to deliver their results. They do so because of their commitment to patients who rely on them to find new medicines.

I am proud to be part of this industry. A medicine given to an individual who is ill changes that person's life. The biotechnology industry is an American strategic national asset, and we have pivoted to meet this national and global challenge. We will defeat this virus, eventually. And when we do, we will reinforce the fact that output of the biotech industry cures patients, can help feeds the hungry, clean our environment, and build a stronger nation. It can do so because the values and intent of the people who work in the industry are rooted in truth, honesty, science, medicine, care for patients, and the constant desire to innovate. They have a purpose. That gives the industry the moral fiber, capability, and intent to tackle what seems unassailable problems.

Biotechnology in the Age of COVID-19 shows how the best of the best have stepped forward in our hour of need and will secure our future. It is truly a remarkable response. I salute my colleagues throughout the industry, some of whose words are recorded in this book, on behalf of myself, my family, my friends, and our nation.

Dr. Jeremy M. Levin is Chairman and CEO of Ovid Therapeutics Inc (NASDAQ:OVID), a company whose mission is to bring treatment to patients with rare neurological conditions. Dr. Levin is concurrently the chairman of the Biotechnology Innovation Organization (BIO). Prior to founding Ovid, Dr. Levin was president and CEO of

Teva Pharmaceutical Industries Ltd., (TLV: TEVA) Before Teva, Dr. Levin was a member of the executive committee Bristol-Myers Squibb Company (NYSE: BMY). In that capacity he was the architect of and implemented the String of Pearls Strategy, which transformed Bristol. Dr. Levin joined BMY from Novartis (SWX: NOVN) where he was global head of strategic alliances. He has served on the board of directors of various public and private biopharmaceutical companies and is currently on the board of directors of Lundbeck (OMX: LUN).

Dr. Levin was voted as one of the twenty-five most influential biotechnology leaders by Fierce Biotech and one of the top three biotechnology CEOs in 2020 by the Healthcare Technology Report. He is the recipient of the Albert Einstein Award for Leadership in Life Sciences and the B'nai B'rith Award for Distinguished Achievement.

He has practiced medicine at university hospitals in England, South Africa, and Switzerland. Dr. Levin earned his bachelor's degree in zoology and a master's degree and doctorate in chromatin structure at the University of Oxford, and thereafter his medical and surgical degrees from the University of Cambridge where he won the Kermode Prize for his work on Captopril.

HISTORICAL PERSPECTIVES

MAKING DRUGS AT THE EDGE OF REASON

Vicki L. Sato

A few days ago, I Zoomed with a Hollywood Squares display of faces on my laptop: the top sixty leaders of VIRBiotechnology, a young company founded with the intent to arm the world against infectious disease. Not a popular vision in 2017, as far as investment was concerned. But on this day, these employees were working across nine time zones to identify new treatments for COVID-19; working from home, working in labs in San Francisco and Bellinzona, Switzerland, to identify and isolate human antibodies that might treat this new pandemic disease.

I am the chair and a cofounder of VIR, and George Scangos, our CEO, had invited me to answer questions from these smart, committed, and stressed leaders.

Early on, someone asked if I had experienced anything like this before. I answered somewhat reflexively: "no, there hasn't been anything of this scope or speed; it's a unique time." Then some older neurons kicked in, and I practically struck my forehead in self-reproach. How could I have forgotten about HIV and AIDS? Today, a generation is unaware of the terror of HIV infection; it is a disease that is "managed" because of medicines that were created in the 1990s and beyond. What lessons did we learn in creating that medical advance?

In 1984 I joined Biogen as a lab rat, an immunologist fussing about interferon and IL-2, but also cognizant of a puzzling and deadly disease—Acquired Immunodeficiency Syndrome—because of my recent academic days reviewing grant applications on a profound immunodeficiency affecting primarily homosexual men. This condition was a tornado, obliterating the immune systems of healthy, young men and killing them within months of diagnosis with opportunistic infections that ravaged their bodies. We didn't know where it came from and we didn't know where it would take us.

We had just learned that this havoc was caused by a new virus, likely from Africa, that infected and killed T lymphocytes, destroying the very physiology designed to protect us from lethal pathogens. Could we imagine anything more diabolical, more deadly, more obscene? Oh . . . and it was a retrovirus, one of the first described with an affinity for human cells, not monkey or chicken, or mouse. Viruses of this type can uniquely embed themselves permanently into the DNA of the host. They can do this because they have an enzyme, reverse transcriptase, that converts the genetic information of the virus, which is RNA, into DNA, the double-stranded form that comprises our own genetic material. They could drive the normal biochemistry of "DNA makes RNA" backward, from RNA to DNA, hence retroviruses. It all seemed rather daunting, and the speed of postgenomic analytical techniques and computer-accelerated modeling, which we take for granted today, were still in our future.

But at Biogen, working on a strategy for a virus-receptor blockade, and later at Vertex working on HIV protease inhibitors, I was able to be part of the scientific *anschlag* that took on the challenge of treating this disease. They were experiences that shaped my outlook on the drama and melodrama of infection.

What did I learn?

Your drug is only as good as your assay

At Biogen, we worked to develop a soluble version of the CD4 receptor, the initial portal for HIV. The idea was to create a decoy that would lure virus away from cells. This engineered molecule worked remarkably well in the in vitro replication assays that measured viral p24 antigen, the gold standard at the time. It did not work so well in patients, though, in part because those early assays employed strains of virus that had adapted to laboratory conditions—strains whose molecular mechanisms had drifted in critical ways from the actual strains infecting people. In addition, we learned that HIV also had co-receptors that figured significantly in viral tropism and entry. Our assays, the most current at the time, had misled us.

Successful and persistent pathogenic viruses are complex

They have evolved efficient genomes to subvert the workings of host cells in ways that enhance their own replication and survival. Single silver bullets rarely work. At Vertex we worked hard to develop a potent, highly selective inhibitor of the viral protease, an enzyme that controlled one essential step in viral replication. It was a powerful drug, but it wasn't enough. Over time, mutational drift and selection created new strains resistant to the drug. Combination strategies, incorporating drugs that targeted multiple steps in viral pathogenesis, were needed to ensure a durable and sustainable response. Inhibitors of viral protease, when added to inhibitors of reverse transcriptase, another protein essential for the virus to multiply, finally tipped the balance toward more durable responses. The combinations were game changers; the individual components were the essential and encouraging steps along the way.

Patients need to be heard and respected

The first generation of drugs were poorly tolerated; the treatment was potentially as bad as the disease, with debilitating side effects and an upside of only weeks to months of prolonged life. Was this enough? Doctors and scientists fretted about doing more careful studies, more innovative studies, before approving the drugs for use. Patients clamored for a voice in that decision, pointing out that their lives were the ones at stake; their lives the ones complicated by handfuls of drugs that caused disfiguring lipid growths, or commandeered their lifestyles with difficult-to-manage, often-contradictory dosing schedules. Maybe weeks or months of life were enough for the moment, if even more effective treatments could emerge in that interval. Patients wanted a seat at the table at the FDA, a seat at the table inside of drug companies, offering insights to teams of scientists. What impressed me so much about this patient engagement was the diligence and sense of community that sustained this effort. Patients who hadn't taken a science class since high school made themselves experts at immune regulation; journalists used to covering crime beats or Broadway shows or financial markets were writing newsletters to keep fellow patients and families informed of the latest discoveries. Patient-activists chained themselves to fences at pharmaceutical companies to demand better access and better medicines. Gay physicians opened neighborhood clinics where patients could find information, advice, and treatments as they emerged, because not all hospitals or physician practices would take on patients infected with HIV. It all made a difference.

Lots of other lessons were learned as well, but these all came rushing back after the figurative forehead slap in front of those leaders at VIR that morning. I believe the lessons are important to us today: know what your assays are telling you and keep making them better. Don't expect a silver bullet cure, but build on steady victories. Be engaged with the breadth of patients affected.

During the time of HIV and AIDS, the biotech industry was in its infancy. It is startling to look back. The only drugs approved from our companies in 1987 were recombinant human insulin, somatotropin, and alpha interferon. Our viability as a set of businesses, let alone as an industry, remained a determined dream for the founders but a "show me" for the rest of the world. We hadn't flexed our muscles yet.

For COVID-19, we aren't newbies anymore. Our industry delivers the majority of game-changing drugs to patients. The speed at which young companies and old have joined hands to test older drugs for rapid efficacy, to launch vaccine trials using a technology only invented a half-dozen years ago, and to map the changing face of this complex virus is breathtaking. We are showing we can do the science and drive the medicine, and that we can invent even as we discover and learn new things about the virus and the multifaceted disease it causes.

But with our maturity comes a heightened responsibility for social equity and responsibility: we must ensure not only that our science is sound but that our medicines are accessible to everyone who needs them. Society is already predicting that we will engage in price gouging, exclusivity, and competition for "credit," the attributes that make our industry one of the least respected globally. Some of our remarkable medicines in other disease areas are better known for their prices than their medical impact.

As we continue to "science the hell" out of this pandemic, let us match our scientific prowess with humility and a commitment to the patients we pride ourselves on serving.

Plagues, it has been noted, have a way of leveling the playing field. Medicines should do that, too.

Vicki L. Sato, PhD, is chair of Denali Therapeutics and VIRBiotechnology. She has spent many years in leadership roles within the biotechnology industry. More recently, she has served on the faculties of Harvard University and Harvard Business School. Sato is also a director of BristolMyersSquibb, Akouos, and BorgWarner Corporation.

VIRBiotechnology dares to imagine a world without infectious disease. It is actively involved in developing treatments and vaccines useful in controlling challenging pathogens.

WHAT THE RESPONSE TO THE AIDS EPIDEMIC CAN TEACH US

Stelios Papadopoulos

On Monday, February 24, 2020, I was the guest speaker at an investor dinner in Boston. Some twenty professionals, among the smartest people in our sector and many with scientific backgrounds, were there. We talked about science, biopharma companies, board governance, drugs, stocks, markets, and, of course, we discussed COVID-19. No one was alarmed; no one thought about getting up to go sit at the other end of the room or to avoid shaking hands. The epidemic appeared to be mostly an academic issue. We debated incidence and fatality rates. We were lulled into believing that, even if infected, the vast majority of individuals without obvious comorbidities would have a disease course similar to that of a bad flu.

Two days later, I was the lunch speaker in another meeting, this one in a Greek restaurant in New York City with some forty Greeks, each one seated tightly next to the other. We shook hands, hugged, kissed each other on both cheeks the way Greek men do, and lingered afterward in small groups. At that meeting, I was to speak broadly about the biopharma industry but not unexpectedly, the issue of COVID-19 came up. My message was—be careful, but don't panic; it's manageable.

Two days later, I was on a plane to Greece to spend the weekend with family and then to Delphi to participate in the Delphi Economic

Forum. When I turned on my phone upon landing in Thessaloniki on Saturday, February 29, the first emails that popped were those relating to the postponement of the forum. Even though no more than a handful of COVID-19 cases had been reported in Greece, the forum organizers thought it wouldn't be prudent to pack 3,000 people in tight quarters. Accordingly, I changed my return flight to New York for Wednesday, March 4.

The night before I was to depart, I received the first of many emails from my colleagues at Biogen alerting me that a high number of individuals among those who had attended a senior management meeting on February 26–27 in Boston were exhibiting flu-like symptoms. Testing was not readily available to provide a definitive diagnosis, but it was reasonable to assume the culprit was SARS-CoV-2.

That's when the epidemic came home to me. In the space of a ten-day period, the epidemic changed from an academic discussion to a personal problem. Week after week, COVID-19 became a personal problem for many Americans and no longer an item on the evening news.

On the plane on the way back from Europe, I played multiple scenarios in my mind, but mostly I was searching for lessons from the past that could provide insights on what was likely to happen. Viral outbreaks of recent years (e.g. SARS, MERS, Ebola) were not useful. The incidence was low and they all had taken place far away from most of us; this was here, it was near us, it was around us, and the numbers were rising at an alarming pace. I realized then that the closest we had been to something like the onset of COVID-19 was the beginning of the AIDS epidemic in the early 1980s. So how are they similar and how are they different? And what lessons can we derive from studying the past?

In the early 1980s I was on the faculty of the Department of Cell Biology at New York University School of Medicine and my wife was a medicine resident at Bellevue, the flagship public hospital in New York City and what became the epicenter of AIDS patient care in the early 1980s. AIDS awareness began through a series of subtle observations by discerning physicians—a cluster of *Pneumocystis carinii* pneumonia here, an increased number of Kaposi's sarcoma cases there. The common thread appeared to be that these patients seemed to experience an unusually severe disease burden, largely because their immune systems were compromised. Another common thread was that the patients were predominantly male homosexuals. In fact, the early name for AIDS was GRID—Gay-Related Immune Deficiency.

Soon the medical observations from different parts of the country coalesced around plausible hypotheses, retrovirologists went hard at work to identify the culprit virus, diagnostic tests were being developed, and drug and vaccine projects got underway.

But in looking back, it is scary how long it took for real medical advances to be achieved in AIDS. There were social, financial, and scientific reasons for that. Society at large viewed AIDS as a lifestyle disease mostly afflicting homosexuals and IV drug abusers (of the 9,600 patients with AIDS in 1985, only 142 had been infected from tainted blood). Consequently, a self-righteous part of society considered the disease to be just punishment for the sinful. Companies were hesitant to embark on drug research in AIDS—there was concern that, for whatever drug or vaccine was to be invented, the government would have stepped in and demanded wide access at a very low price. But the biggest challenge was the state of scientific capabilities measured against the degree of difficulty of the problem at hand. All these factors conspired to slow progress and, in fact, if it had not been for the efforts of the AIDS activist community, much less would have been accomplished.

The enormous difference between our response to COVID-19 compared with AIDS is the ferocity with which the biopharma sector has committed resources toward solving the problem and the extraordinary progress that has been made in a short time. Yes, science is more advanced today than it was in the early 1980s, but importantly we have a far more dynamic, resource-rich, and morally responsible biopharma industry—a big part of what motivates many of us to work hard in improving the human condition is our sense of duty.

Consider the rate of progress. From 1980–81, when the first AIDS cases were reported in the US, it took until March 2, 1985, for the first diagnostic test, an ELISA-based assay for the purpose of screening blood, to be approved by the Food and Drug Administration (FDA). To ensure the safety of the blood supply, the test was highly sensitive. Consequently, when used to test patient blood for the purpose of offering a diagnosis (for which it was not approved), it resulted in too many false positives. The first rapid HIV diagnostic kit was approved on November 7, 2002, and the first at-home diagnostic kit was approved on July 3, 2012. By comparison, for SARS-CoV-2, under FDA's Emergency Use Authorization, we already have more than one hundred PCR-based virus detection kits as well as serological tests to detect antibodies in patient blood. True, the tests are not as accurate or scalable as we would like them to be, and an antigen test is needed so we can proceed with massive screening, but the progress is still remarkable.

On March 19, 1987, zidovudine (AZT), a compound originally synthesized as a potential anticancer agent, became the first drug approved by FDA to treat AIDS patients. As of the end of April 2020, less than six months since the first reports of COVID-19 from China, more than 300 clinical studies are focusing on repurposed drugs, direct antivirals already studied against similar viruses, novel antibodies, and convalescent plasma. It is very likely that over the next six to twenty-four months, one or more of these studies will form the basis for the approval of agents that could help improve patient outcomes.

In addition, multiple approaches are deployed for novel vaccine development. Here we would have to be a bit more cautious in our expectations compared to the dialog in media and the general public. It appears that we state axiomatically that we will have a vaccine, and the only question is whether it is six months or twenty-four months away. A vaccine against a novel agent and based on mostly novel platforms is complicated, and it may well be many years before a truly efficacious and safe vaccine is developed. A sobering reminder is that we are in our fourth decade of trying and still do not have an HIV vaccine.

How did the world cope with AIDS during all those years when not much was available for treatment? The panic subsided when we came to understand the biology of transmission. Early in the epidemic the fear of the unknown prevailed. People were concerned about their food being prepared by infected individuals, shaking hands, or being in the same room with them. Medical staff would double-gown and double-glove before entering the rooms of AIDS patients. And the burden to the system was overwhelming. At one point in 1985, a quarter of the 200 medicine beds at Bellevue were occupied by AIDS patients mostly kept in isolation, one patient per room. Worst of all was the fear of accidental pricking by a needle used on an AIDS patient. As it became clear that AIDS patients were more at risk of getting sick from healthy people because of their severely compromised immune systems, rather than infecting those around them through casual contact, the pressure was lifted from society.

The lesson is that it is of paramount importance to understand the biology of transmission of SARS-CoV-2 so we treat those who require treatment and redirect resources away from those who will navigate benignly through a potential infection or are not likely to get infected. The scientific community is well aware of these issues and that what stands on the way of answering the transmission-related question is mostly resources—reliable and scalable diagnostic tests so we can engage in massive testing. In addition, we need the self-discipline to carry out such studies with rigor, sacrificing time in order to end up with scientifically defensible answers.

The current pandemic has had a deleterious effect on economies, and the worst is probably still to come. Restarting economies and redeploying the extremely high number of unemployed will be a challenge, particularly because many of those out of work now will never be able to return to their old jobs, given that many professions and businesses will be redefined. And the crisis will be exacerbated when the virus spreads over the next few months to developing nations, particularly in Africa, where there is limited opportunity for social distancing or tertiary medical care.

The past few months have been a trying time for people in this and many other countries. But not everyone in the US has been affected the same way. Affluent families shelter in place in comfortable suburban homes or weekend retreats, they get food and goods delivered to their doorsteps, and if they get sick, they have ready access to quality health care. Inner-city middle- and lower-middle-class families, now out of work and in many instances virtually insolvent, are trapped in crowded apartments and are far more likely to get infected. Once they fall ill, they end up in overcrowded and under-resourced city hospitals or die at home. The income inequality, already recognized as a significant social issue in this country, has now become a survival inequality. Along with new drugs and vaccines to fight the virus, we need to rethink our health care delivery system to ensure adequate access to all.

Dr. Stelios Papadopoulos is chair of Biogen Inc., Exelixis Inc., and Regulus Therapeutics Inc. In the not-for-profit sector, Papadopoulos is a member of the Board of Visitors of Duke Medicine, a member of the Global Advisory Board of the Duke Institute for Health Innovation, and cofounder and chair of Fondation Santé, a foundation providing research support to biomedical scientists in Greece and Cyprus.

IT'S THE SCIENCE, STUPID

Samuel Waksal

So much about the COVID-19 crisis currently ravaging the world and causing a global pandemic is new. We call this virus a "novel coronavirus" precisely because humans have never encountered it.

This virus and a number of others like HIV, influenza, Ebola, and SARS have come from animals. After infecting humans, they mutate and pass from one human to another. Some, like Ebola, are so deadly that they cannot truly create a global pandemic.

Others, like influenza, spread more rapidly but are not nearly as deadly.

Every pandemic presents a new case to the scientific world and into to the world in general. As Adam Kucharski says in his book *Rules of Contagion*, those who model pandemics like to say, "If you've seen one pandemic, you've seen . . . one pandemic."

The urgent challenge for the world scientific community and the biopharmaceutical industry is to figure out how to respond properly and how to end this global search scourge.

We all yearn for a better sense of when "normal" will return. We all want to see the light at the end of the tunnel. The economic, psychological, and interpersonal tolls of this crisis are profound. It has closed down a good deal of the world. So this is much more than an infectious disease. It has taxed the world in ways we could not imagine, in both the developed world and developing nations. It has taxed our supply chain capabilities and our health care institutions. It has taxed our physicians and nurses. And it has taxed our thinking about preparedness.

Because we are in such uncharted, profound, and dangerous waters, the world is looking to the scientific and biopharmaceutical communities to get us out of this crisis. This is ironic.

Over recent decades, much of the world has looked at science with considerable suspicion, questioning the very truths on which science is based. Yes, science at its core deals with discovering truths. The people who ignore these truths attack science and deny the essence of what has extended human life for generations. They are against vaccines and the industry that makes them. They don't believe vaccines work. "Anti-vaxxers" prefer myths over scientific truth.

At the same time, the biopharmaceutical industry, which uses scientific truths to generate cures for diseases that attack mankind, has been attacked as only pursuing profits. In the United States and Europe, the funding of much of the important medical research that is generating novel drugs and diagnostics has also been attacked. The ability of biopharmaceutical companies to price innovative lifesaving drugs, has been attacked as well.

For example when Gilead developed a cure for hepatitis C infections, they were attacked for charging too much money for their drug. The fact that it saved society billions in future costs of liver transplants and treatment of liver cancer in patients with HCV was ignored.

This is the same company that people now look to for manufacturing remdesivir as a therapeutic intervention for COVID-19. Gilead spent tens of millions of dollars developing this drug which is now helping people recover more quickly from COVID-19, is saving lives, and preventing hospitals from being overwhelmed

So, when a global pandemic endangers the ability of the world to function normally, what will save us? Only science can because it has the ability to find answers to the unknown enemies of mankind.

Over the years, the understanding of microbes both bacterial and viral change the way we think about infectious disease. They have helped the world create a generation of vaccines that have prevented the deaths of millions of people in the world. And it has been done by the biopharmaceutical industry.

In 1854,John Snow, a young English Doctor, published his new ideas on germ theory focused on the mode of communication of cholera. All the deaths from cholera were close to the Broad Street pump in London. Because of his unique view of the mechanism of disease transmission, he searched and found the well that was the water source of all the cholera cases he was investigating. Thus, Snow became the founding father of epidemiology. His search for scientific truth disproved the dominant theory, that plagues were caused by pollution or a noxious form of bad air, "miasma."

During the 1918–19 influenza pandemic that killed over fifty million, around the world, scientists responded in ways they never had before. And because of that global pandemic, scientific institutions leaped into modern times, especially in the United States. Scientists had learned how to culture the polio virus and began to work on a vaccine, which about thirty-five years later became a savior to generations that followed. American medical schools and research centers like Rockefeller Institute (now Rockefeller University) became model centers for the world and the places for some of the greatest modern discoveries.

Oswald Avery was one of the young scientists searching for the etiologic agent that caused the great influenza. He later discovered that DNA was the molecule of genetic material and that later led to the discovery of the double helix molecule by James Watson and Francis Crick.

In the 1940s, because of the discovery and synthesis of penicillin at Oxford, England, the world was changed. But it was a small chemical company in Brooklyn, New York, called Pfizer that manufactured it. Penicillin was one of the drugs that changed the lifespan of humankind. Pfizer is now one of the world's great pharmaceutical companies working on a COVID-19 vaccine.

As with the cholera epidemics of the mid nineteenth century, the great influenza of the early part of the twentieth century, vaccines for polio, penicillin for bacterial infections, and the antivirals for the HIV epidemic in the latter part of the twentieth century—science changed the way we understood these diseases.

We learned to use scientific research, medical science, and medical discoveries to alter outcomes with new treatments for many diseases.

During the HIV epidemic, America's research centers and biotechnology companies discovered and sequenced that emerging virus very quickly. Once that was done, many companies were able to create novel HIV therapeutics. Dr. David Ho suggested that combinations of direct-acting antivirals would be the way to treat HIV infections. Because of that, companies like Merck and GSK were able to make a new generation of antivirals, and HIV/AIDS was converted from a killer to a chronic disease. Gilead later combined these different antivirals into a single pill to make treatment easier.

These events that occur during epidemics and pandemics change how we think about innovation and science.

Our industry has always been a marriage between academic research institutions and the biopharmaceutical companies that move interesting discoveries forward. We are watching that happen now.

The biopharmaceutical industry has also been working on creating a new generation of antiviral therapeutics as well as vaccines for the betterment of humankind. Cetus was the biotech company that gave us the PCR test, which was later developed by Chiron, Roche, and Boehringer Mannheim. It changed the way we do direct-acting tests for viruses like this novel coronavirus. It is the basis of the diagnostic test for COVID-19. A scientist working in a biopharmaceutical company that invented the PCR assay, and helped us change the way the world worked. For his pathbreaking efforts he was awarded the Nobel Prize for chemistry. And yet the same industry has been under attach fr creating the innovation necessary to return us to some kind of normal 2.0.

Oxford University is working with Astra Zeneca to try to move a new vaccine candidate forward. Abbott laboratories and Roche are helping with new diagnostics to be able to better determine who has been exposed to this virus and who has developed immunity.

Jansen Pharmaceuticals, part of Johnson and Johnson, is also working on an exciting new vaccine using the spike protein of COVID-19. All of the biopharmaceutical companies that are working on new vaccines are moving things forward in record time to help society get through this pandemic.

Dozens of other biopharmaceutical companies, large and small, are working on both vaccines and therapeutics to help us get through this pandemic. They are developing new drugs and vaccines all based on scientific truths. They are also helping us deal with fears of the unknown that have affected our lives in a multitude of other ways.

And the development time frames are getting much shorter. This is a pandemic which began in China towards the end of 2019. It raced through parts of America in February and March 2020. Already by May, our biopharmaceutical industry was making monoclonal antibodies as therapeutic modalities for the treatment of COVID-19. Meanwhile, scores of other biopharmaceutical companies are working on a new generation of vaccines for this virus. What used to take years is now taking weeks.

Our current approaches to testing have missed many of those who have contracted the virus but cleared it. As a result, the publicized numbers for infections in our country and globally are only the tip of the iceberg. The real numbers now appear to be much higher. Already we are seeing that 20 to 30 percent of the people in places like New York have been infected and are probably immune. That number is growing as testing becomes more widely available. And so our immune systems are in a race with the biopharmaceutical industry to win the war against this pandemic.

Governments are working hand in hand with the biopharmaceutical industry to generate the newer data which will provide us all with a more accurate assessment of the course of the pandemic. This is a pandemic and it takes a lot of partners.

As more data comes in, we learn more about this novel virus. And that data will provide the scientific basis for establishing how to combat it and how to ensure that the future is more secure for all mankind. As we learn how to counteract COVID-19, scientists in the biopharmaceutical industry can also better prepare us for dealing more effectively with the next novel virus. One that could pose an even greater threat to our societies, one that will require the rapid development of a new generation of antivirals and vaccines. If we allow science to prevail, our lives, and our world, will be much better for it.

We know that human memory evolved to help us predict the future. Now we have to use the science and what we have learned about the past behavior of viruses, including this coronavirus, to better predict its future course.

All the while scientists both in academia and the biopharmaceutical industry continue to work on innovative drugs and vaccines as well as cellular and gene therapies for a host of other diseases which affect humankind.

The biopharmaceutical industry is not just doing it for infectious diseases. It has altered outcomes for cancer patients, for children with genetic diseases, and people with genetic forms of blindness, and it is beginning to change the world for neurological diseases. And this is being done by companies doing important and innovative work like Meira Gene Therapy.

I hope that this pandemic will teach people not to heed those who use soundbites to stoke animus against science and the industries science has spawned. I hope we come to understand that scientific innovation and scientific truth is what changes the world. It doesn't change the world unless we in the biopharmaceutical world are encouraged to innovate and create.

When Bill Clinton ran for president, one rallying cry was "it's the economy, stupid."

What we are learning from this pandemic is that "it's the science, stupid."

Dr. Samuel Waksal has devoted his life to building notable companies based on innovative science. He was the founder and former President and CEO of ImClone Systems, developer of Erbitux, one of the top-selling cancer drugs in the world, as well as Cyramza, the first anti-VEGFR-2 antibody. Dr. Waksal also founded Meira GTx an international gene therapy company. In addition, he founded Equilibre Neuroscience, Kadmon Corporation, and Merlin. He serves as a visiting fellow in neuroscience at Weil Cornell Medical College, Feil Family Brain, and Mind Research Institute in New York City.

BENEFITS OF SCALE: HOW BIOPHARMA SHOWS UP IN A CRISIS

Bill Sibold

When asked "what do you do?" I know some colleagues who sheepishly reply, "I work for a big pharma" or "I work in biotech." Both "big pharma" and "small biotech" have their own perceptions and misperceptions, from the public and even from each other. I lead the biotech division of a big pharma company, so I have the opportunity to experience both and know that the innovation and reach of both will be needed to solve this crisis.

I have worked in the biopharma industry my entire career, and say every day that it is the best industry in the world. Why? We solve problems, which in this industry means that we transform and save lives. We seek out the world's biggest medical challenges, and we figure them out. And this has never been more obvious to me as we battle the pandemic of COVID-19.

I know what we're all going through is new to us and feels like it's never going to end. But we need to take the long view. We've had pandemics before, not to mention countless national and global disasters and challenges. We have overcome them every time. We figure it out; we roll the stone up the hill no matter how many times it feels like it's about to roll back on to us.

Throughout history, we have had to battle similar hardships with contagious viruses. The Spanish Flu pandemic in 1918, the deadliest pandemic in human history, killed an estimated twenty to fifty million people worldwide. People were ordered to wear masks and schools, theaters, and businesses were shuttered as the world waited for the virus to run its course. With no known treatment or vaccine, the human race was powerless to stop it.

Through scientific discovery and perseverance, the first influenza vaccine was discovered in an academic lab in the 1940s. However, innovation is successful only if it can be implemented. Innovation alone can't conduct massive trials nor produce and distribute a vaccine on a large scale. It was the assistance of pharmaceutical companies that helped deliver this innovation to the world.

Recently, I felt an interesting déjà vu with the numerous announcements of biopharma companies' collaborating with each other and academic institutions to quickly develop and scale the production of promising vaccine candidates for COVID-19. Collaborating and scaling important scientific discoveries has been the backbone of this global industry from the beginning. I am glad to see this situation may be able to benefit from this approach.

We speak about innovation a lot, and we should, as it's the beating heart of our industry. But innovation alone will not meet this challenge. The spark of innovation hits the reality of scale, distribution, global regulatory issues, and every other complexity that Sanofi manages daily.

Translating ideas to reality needs the ecosystem of innovation. Sanofi has each part of this ecosystem under one roof and in moments like these can quickly call upon the full resources of the company. We can draw upon a rich library of therapeutic areas and over 100 years of history and heritage. Whether it's looking at older drugs that may have new applications, or looking at our currently marketed biologics in immunology, we have the bandwidth and resources to explore every opportunity.

If you needed surgery, would the doctor come into the operating room with just a stethoscope? No, they bring every tool needed to fix the problem. This is what biopharma is for this pandemic. We don't have one tool; we don't focus on one sliver of research, development, or distribution. We can discover, develop, produce, and deliver millions of treatments to every country that needs treatment. We know this because at Sanofi we do this every day.

When I look at the fast and agile action that we at Sanofi have undertaken in a short time, I am gratified by the possibility of what we will be able to achieve. We are the only company in the world working on two vaccines, a self-diagnostic, and two potential treatments for this highly contagious and deadly virus. And, remember, we haven't ceased any of our other operations. Every patient who counted on us before counts on us today. Their rare disease, their asthma, their multiple sclerosis doesn't take a break, and neither can we.

This is what makes biopharma our best shot at success. We can try, and fail, dust ourselves off, and try again. We will figure this out. We specialize in innovation, but just as important, we can then deliver innovation at a global scale, from the lab bench to every country, city, village, and family that needs it. This is a global problem that requires global thinking and a global solution.

When this is over, will anyone recognize how much biopharma stepped up to avert disaster? Frankly, I don't know. But when asked "what do you do?" every one of us who works in this industry should say "I help to transform and save lives." It's what we do.

Bill Sibold, executive vice president and head of Sanofi Genzyme, has more than twenty-five years of experience in the biopharmaceutical industry since starting his career with Eli Lilly. He has held leadership positions at Biogen and Avanir Pharmaceuticals. Currently, Bill is a member of the Sanofi executive committee.

Bill holds an MBA from Harvard Business School and a BA in Molecular Biophysics and Biochemistry from Yale University.

Q&A WITH A BIOTECH PIONEER

Frederick Frank

Frederick Frank has been my friend and mentor for over thirty years. His voice is known throughout the industry, and he played a pivotal role in forming the biotechnology industry as we know it today. I interviewed Fred and it's an honor to have his thoughts.—JEREMY LEVIN

Tell me about your background and what brought you into the industry.

After serving in the Army for two years during the Korean Conflict, I went to the Stanford Business School, thanks to the GI Bill. Since I had no idea what I wanted to do for a career, I hoped that in business school I would find the spark for my future direction. And, indeed, that was the case. I set out on a career in investment banking, joining Smith Barney in their No. 1-ranked research department. I was assigned to work with the head of the department, who was also the drug and chemical analyst. After working for him for six or seven months, I said to him that it makes no sense for someone to cover the drug and chemical industries; they are not related in any meaningful investment way. I suggested that he cover the chemical industry, as he was a chemist by training. He agreed, and voilà, I became Wall Street's first dedicated drug analyst. Five years later, I became the first pharmaceutical industry investment banker. Sixty years on, COVID-19 struck.

You saw the dawn of biotech and played a critical role in its inception and evolution.

In 1978, Dr. Alexandro Zaffaroni, the chair and CEO of ALZA, called me. He asked me to come to San Francisco to visit a new company he joined as a director. So off I went to Berkeley to meet with Cetus, the pioneering biotech company. Learning about this new world of molecular biology was truly eye-opening and fascinating. I started visiting Genentech about a year later. And so, my deep involvement in this new area began. At Smith Barney, then later at Lehman Brothers, which I joined as a partner in 1969, I was fortunate to be on the frontier of an exciting new chapter in the life sciences industry.

How did the traditional pharmaceutical industry view this new segment of the life sciences industry?

The pharmaceutical industry was both skeptical and dismissive of these "new life sciences companies."

What changed that attitude?

The dramatic change occurred when I engineered a first-ever broad-based collaboration between Roche and Genentech. The deal was announced, and the industry changed. A pharmaceutical giant committed to biotechnology. The sequence of the Genentech-Roche collaboration over time resulted in Roche's acquiring Genentech for an aggregate consideration of over $46 billion. This ushered in an era of intense collaboration between the pharmaceutical industry and the new biotechnology industry.

What was the outcome of that intense interaction?

The rise of the biotech industry is a phenomenal chapter in both the life sciences industry and corporate America. Today the relationship is symbiotic. Indeed, over 70 percent of all new drugs in the industry originate from biotechnology companies, and billions of dollars of research and development monies are invested in the biotechnology industry by the large global pharmaceutical companies.

The press and the public often criticize the industry. As we entered 2020, the administration was poised to pass legislation to mandate controls and other types of legislative restrictions to reduce drug prices. What brought us to this situation?

By far the most contentious consideration underlying the hostility of the various constituent groups, insurance companies, government payers, and patients is Rx prices. And an important factor in this pronounced hostility is the pace of drug price increases and the rate of price increases. Many companies have increased the prices of their leading drugs twice a year and at rates far above underlying inflation. It is difficult for the drug companies to justify the substantial price increases that have gone on for a number of years. Part of the explanation given by the drug companies is that the discounts they are forced to provide the payers underlie their "need" to increase their prices so substantially. Virtually no one buys into this argument. The fact that the major companies have the highest profit margins of virtually any industry and the highest return on equity is not lost on the industry critics.

COVID-19 struck China on November 17, 2019. By early January 2020, our industry had begun to respond. What is your impression of that response?

Clearly, the fact that the world is having to deal with a pandemic is having multiple repercussions. And quite clearly, the working masses are fed up with being out of work and small business owners are equally frustrated. The financial burdens on so many earners have become intolerable. The likely repercussion is the risk of increasing the pandemic in many regions as the pressure to open businesses increases. This pressure to "get back to work" will itself reach epic proportions. Unfortunately, in my opinion, an effective vaccine is further out than we are being led to believe. All in all, an intolerable situation for the administration, state governors, local mayors, and the public. Many have reached the point where they are willing to take the risk of getting the infection in order to get back to work. Quite ironic that so many who complained about work now relish the opportunity to get back to earning a paycheck.

What could we have done better as a nation and as an industry?

Wow, a very difficult question. The contentious political divide is a major hurdle to getting effective programs in place. Look how severely the president was criticized for his closing access to the country by the Chinese. Even though that was a bold and appropriate decision, he still gets criticized for doing so. Clearly, what we could have done better is to follow the Boy Scout motto, "Be Prepared." But forward thinking, unfortunately, is not in the DNA of our politicians or the populace at large.

What do you think the lasting impact of COVID-19 will be on our industry and on our nation?

Again, a very difficult question as the pandemic is still playing out. What is clear is that the past is not prologue. The repercussions on our industry are both positive and negative. While there will likely be less criticism and harsh condemnation of the drug and bio companies, the issue of drug pricing and what to do about it will remain a heated political subject. The fallout on the nation at large is also likely to change the landscape dramatically. There are broad implications for all aspects of society. Retail establishments will be severely compromised. Online retailing will continue to garner a larger slice of the pie. Consumers are likely to be more judicious in their spending habits. In short, no aspect of pre-pandemic life and everyday habits will be untouched.

And in the wishful thinking category, I hope society at large will become more civilized, respectful of one another, of the police, teachers, religious groups, the military, virtually all constituent groups. I have observed in recent years that we call modern life "civilized" where, in fact, it is far from being civilized. How wonderful if civilized behavior became the operative way of life.

Do you have any last comments?

It is truly ironic that the sad emergence of COVID-19 has transformed the drug and biotech companies from being the bad guys (black hats) to the good guys as we try to contain and defeat this scurrilous, worldwide epidemic. It proves the fundamental raison d'être for these companies, namely, to improve the quality and longevity of life.

Frederick Frank is an investment banker with over sixty years of experience on Wall Street. He is credited as lead underwriter for over 125 IPOs and as a negotiator in over seventy-five mergers and acquisitions. Frank was chosen as one of the Top 100 Living Contributors to Biotechnology in 2005. .

BIOTECH TO THE RESCUE, AGAIN!

Brook Byers

Humanity relies on the thousands and variety of biotech companies (and their larger pharmaceutical partners) to save lives and reduce suffering from pathogens, heart disease, cancer, inflammatory and CNS conditions, and the 2,000 human genetic diseases. In my forty years of cofounding, financing, collaborating, and building diagnostics and drug companies based on biotechnologies, I've seen variations of appreciation, awe, support, criticism, and confusing regulation of this essential industry.

Perhaps a "silver lining" of the COVID-19 pandemic is the amplified awareness by government officials and the public that the solutions in diagnostic testing, drug treatments, and vaccines will come from this industry as it translates basic science discoveries into safe and effective products. The biotech industry deserves more support, such as strong market incentives (versus penalties), modernized regulation, higher diagnostics reimbursement to incent innovation, and consistent federal and state purchasing to prepare and stockpile against the most likely pathogenic pathogens for the inevitable next pandemic.

In 2006, Congress passed legislation to establish the Biomedical Advanced Research and Development Agency (BARDA) division of Health and Human Services, modeled after the Defense Advanced Research Projects Agency. Its mission is to identify, fund, and assist breakthrough projects to prepare for and manage threats such as pandemic influenzas and emerging infectious diseases, plus other challenges. BARDA to date has supported the development of forty-two FDA approval products addressing such threats. I am proud to have been on a team working with senators, House members, and the White House in 2005–2006, who saw the need to innovate and be prepared. We need leadership that understands and consistently and aggressively supports this agency as well as other pandemic preparedness federal and state programs.

Given the trillions of dollars spent to shore up the American economy and health care system because of COVID-19 in 2020, why not spend more to prepare in advance and avoid being caught behind next time? The US government subsidizes the fossil fuel industry for $20 billion per year. Why not spend more on biotechnology research, pandemic preparedness, and incentives for the biotech industry?

Science over politics, please!

Brook Byers is a Partner at Kleiner Perkins, a forty-eight-year-old venture capital firm in Menlo Park, California, which has helped entrepreneurs build over 600 companies. Over 100 of those have been science-based-biotech and medical ventures with FDA approved products that have been used to diagnose and treat many of millions of patients.

SCIENCE IN THE SERVICE OF HUMANITY

Sol Barer

As I contemplate the pandemic, I recognize that, by virtue of age and prognostic factors, I am at significant risk. There's no way to avoid it, so at times like this I contemplate whence I came and my path—indeed I become a bit philosophical. We arrived in the US on the SS *MacRae*—a ship converted by the Navy to transport displaced persons from Europe after World War II. When I was ten, the Soviet Union launched the first satellite, Sputnik—and that changed the perception of science by many in this country. The US needed to catch up, and initiatives to promote science and science education arose; as someone from a poor family, it offered me a path to a profession. Plus, I liked science. Even more importantly, my mother wanted me to be a scientist. My perspective on "science in the service of humanity" took shape—and has guided me throughout my life. Science was eminently logical, disciplined, and ultimately yielded wonderful results. It was hard work, but nevertheless would lead to advances in all fields and elevate humanity. Surely everyone could see this.

I was fortunate in my choice of career and achieved some success with Celgene, where I was surrounded by bright, committed people with high ideals. Science in the service of humanity—"doing well by doing good."

Then I grew up. Much of what I believed about the perception of science was not universally shared. Conspiracy theories and ignorance were ever-present. I despaired of this public perception and our ability to effectively harness my beloved science in the face of everything from "anti-vaxxers" to "flat earthers" to conspiracy theorists to disdain for my industry, biopharma.

Then came the pandemic. I watch science arise, and especially the biopharma industry come together and focus in unprecedented ways. The enormous and prodigious intellectual power of the people in the industry focused on one objective—the end of the pandemic. The decades of research by industry, by academia, by governmental researchers formed a base for a mobilization never seen in the biological sciences—indeed perhaps to rival the great efforts of the Manhattan Project and the moon program. In an incredibly short time—a few months—several vaccines are in development, promising therapeutics are being developed, regulatory and governmental agencies are responding in an unprecedented manner. Biopharma has mobilized and it is an awesome force—working together. Therapeutic development at this pace hasn't been seen before.

And the public? Scientists are hailed as heroes, scientific leaders are the most respected, health care workers are idolized. As my wife and I sit isolated in our home, the world has changed. Young people talk about public health and science and medicine careers.

Will this last? The costs of the pandemic have been profound—lost lives, economic hardship, and social disruption. Yet there will be untold benefits: from effective interorganizational practices and profound lessons in the nature of leadership to the advances transcending the specifics of the pandemic to many medical areas. Indeed, the multitude of therapeutics developed will be awesome. The true value of the scientific endeavor has become clear to so many people. The promise of "science in the service of humanity" indeed is a visible mantra.

Sol Barer received his graduate degrees in organic chemistry and was a founder of Celgene, where he retired as chair and CEO. He is an adviser and board member of numerous companies and not-for profit entities, including biotechnology companies, and is chair of the largest generic company, Teva Pharmaceuticals. Barer feels strongly about the role of the biopharma industry to significantly advance medical care—it drives him and the companies he is involved in. He believes this is what makes our industry special and valued.

PERSONAL PERSPECTIVES

A PANDEMIC, A RESET, A FUNERAL, AND A CHANCE TO HEAL THE WORLD

Michelle McMurry-Heath

Cell phones started ringing at a board meeting of the National Patient-Centered Clinical Research Network. Concerned faces and furrowed brows signaled the gravity of the news. Janet Woodcock, the director of drug evaluation and research for the US Food and Drug Administration, stood up and excused herself.

A few weeks earlier, at the J.P. Morgan conference for biotech investors in San Francisco, I was huddling with my colleagues on Johnson & Johnson's external global leadership team when our executives from Shanghai and Australia dished scuttlebutt about a virus disrupting J&J's Asia operations. We were later shocked that two full weeks of Chinese New Year celebrations, normally sacrosanct, had been canceled outright by the government. Another SARS epidemic, I told myself: a serious but regional outbreak.

Then, foreboding news closer to home suggested otherwise. An Episcopal priest in the nation's capital tested positive for the novel coronavirus. Concern quickly spread through the congregation that parishioners were becoming sick—some, very sick. My daughter Isabella's elementary school, part of the National Cathedral and the Episcopal diocese, abruptly canceled classes four days before their scheduled spring break.

Still, I remembered how the school had shut down for a week last Thanksgiving because of a norovirus scare. Maybe this was like that. Maybe the entire school year wouldn't soon be canceled. Maybe hundreds of thousands of lives wouldn't soon be lost. Maybe a quarter of the country wouldn't soon be out of work. Maybe the worst global pandemic in a century wouldn't soon be actually upon us. Maybe . . .

It's interesting the negotiating we do with ourselves when we're processing a paradigm-shifting catastrophe. As a pediatric resident at Children's National Medical Center on 9/11, I remember watching from a hospital breakroom as the Pentagon exploded. The hospital went on lockdown as we tried to figure out if the Pentagon's day care center had been hit. (Thank God, it hadn't.) And then we dealt with huge lines in the ER a few weeks later when the anthrax attacks happened. The National Capital postal workers waited hours—not to see if they were affected, but to be certain their children were not. It was all so surreal. At the hospital, parents became more likely to freak out, yelling at doctors and staff. Being a young doctor and not yet a parent myself, I had trouble understanding their reaction. Now with an eight-year-old, it is obvious. There is nothing more humbling than feeling powerless to protect your children.

As the scope and lethality of COVID-19 became apparent, my internal negotiator balanced my desire to be a responsible public health leader with my maternal impulse to worry and protect. I tried to rely on my scientific knowledge and not overreact. I didn't want to scare Isabella. But she, too, soon realized the coronavirus was not like the norovirus when we were forced to cancel our spring break ski trip twelve hours before our flight's scheduled departure. They had closed the slopes.

I explained to Isabella that the virus did not seem to hit children as hard. And I told her we would do everything in our power as a family to stay safe. Still, she struggled to understand why she couldn't see her friends with whom she had played just days before. Isabella had some nightmares. So did her mother. Truthfully, she is still processing the fear and probably will for some time to come.

COVID grounded me and brought me home, after years of sometimes-grueling world travel across China, Singapore, South America, and Europe. It made me reflect on the importance of family and community. Right on time, my wonderful husband, Sebastian, a large-animal veterinarian, had a eureka moment: Let's make fruit care packages for seniors in our Forest Hill neighborhood. I knew that something about these small acts of kindness would reverberate.

We bought and arranged the healthy produce. Isabella prepared a pun to go in each package: Why didn't the apple understand the orange? Because he didn't speak Mandarin. We first visited the widow of a prominent doctor, discovering she had returned home from the hospital that day. She had no one to care for her, so we did. We went to another house up the street. The gentleman was so happy to see us. His wife, too, had recently returned from the hospital. She had respiratory problems, and the paramedics made her sit on the front porch for hours of testing because they thought she had COVID. She didn't. She just had respiratory issues and bad timing. The fear was palpable. But so was the sense of community. Despite living in our neighborhood for nearly ten years, we were taking the time to get to know our neighbors in a new and profound way.

I knew what that fear felt like because my college sweetheart and first husband, Peter, was diagnosed with cystic fibrosis at age four. He had spent so much of his twenties and thirties working harder than you can imagine staying healthy enough to preserve his vital organs while scientists devised a cure. Peter lived through some of the cruelest contradictions of modern medical research. Although we were taught in medical school that double-blind, placebo-controlled studies are the gold standard of medical research, no one ever talks about how hard it is to be a subject in one of those studies. Peter was running through his precious lung capacity when a new, promising therapy started to emerge. He entered the trial, but we were terrified that he might be inhaling placebo every day.

Luckily, he started feeling much better. When the trial was unblinded, we discovered that he had received the real drug and the results of the study showed that it really was a breakthrough. But I still wonder what would have happened if he had just received saline instead. There was growing interest in cystic fibrosis, but that interest was dwarfed by interest in diseases that could mobilize celebrities and activists. This central premise—your access to the scientific, intellectual, and fiscal capital required to fight any illness should not depend on demographics—has shaped my entire career since.

Peter inspired me to study the political and financial levers that promoted or obstructed medical progress for the people I grew up with in Oakland, California. I wanted to know why so few of them would ever gain access to the latest and greatest care. How do we incentivize scientists to address the questions that matter to patients and make sure that the progress they make reaches the neediest patients? We talk a lot about access to health care in our country, and rightfully so. But we also need to talk about access to science. The best coverage won't buy you better health if innovators haven't worked on an effective treatment for what ails you. Doctors need tools and, without them, they are powerless. Peter inspired me to write legislation as a health policy aide to Senator Joe Lieberman directing the National Institutes of Health to providing more research funding for diseases with smaller or poorer patient populations or diseases where the science was too challenging for most to tackle. A version of this bill was enacted into law as part of the Affordable Care Act in 2009.

One in ten Americans suffers from one of 7,000 rare diseases; only 5 percent of those diseases have approved FDA treatments. Having a disease for which there is no cure is about the most tenuous, frightening way you can imagine living a life. Now, because of COVID-19, everyone in the world understands that fear—the fear of contracting a disease for which there's no cure. It's a psychological sea change.

Peter, like so many cystic fibrosis patients, eventually needed a lung transplant. Then, last December, like so many CF patients with late-stage disease, he died. He was born too soon for the new miracle CF therapies. Sebastian, Isabella, and I flew to Ithaca in January for his memorial service. His death, poignant and tragic, crystallized for me how many people die waiting on solutions. And how many lives are touched by their loss.

Three days after we returned to Washington from Peter's funeral, this email from a headhunter popped in my inbox:

"I wanted to connect with you regarding the search we are just launching to recruit Jim Greenwood's successor as the CEO of BIO. Given the association's focus on the innovative, patient, and societally oriented work of its members, we're looking for an energetic and visionary leader with industry experience, an understanding of science or medicine, and the ability to serve as a visible spokesperson and advocate for the important work being done in the biotechnology space."

I've worked as a physician, a molecular immunologist, a science policy advocate, an FDA official, and an executive at a large pharmaceutical company. But I've always known deep down that the most effective use of my skills is to be an advocate for science and scientists, writ large. That has always been my passion: getting new medicine and new hope to patients.

As the interview process unfolded right before the quarantine began, I sat down with BIO's current and future chair in New York. I remember saying to them: "If you're looking for a typical lobbyist, I'm not that person. I'm an advocate for patients and the science we can bring to them. And that's how I'd approach this role. I'm not here to package whatever bad-smelling position some people want to sell to make it palatable. That's not what I do. I'm a true believer, and if I believe in something, I'll fight for it to my dying breath."

This spring, I received the word that the board of BIO, the Biotechnology Innovation Organization, unanimously voted for me. I was truly elated to receive the news. Every career experience in my life has prepared me for this moment. BIO has the unique ability to be an honest broker with government institutions and biotech companies across the world so we can avert suffering and save lives. Working side by side, we can help deliver COVID diagnostics, therapeutics, and vaccines to patients in every corner of the globe.

Biotech entrepreneurs and scientists are about to set the world on fire. I couldn't be prouder for this opportunity to be their champion.

Dr. Michelle McMurry-Heath became just the third BIO President and CEO on June 1. 2020, succeeding former Rep. Jim Greenwood. She is a physician-scientist with an MD/PhD in molecular immunology. She led a global regulatory team of 900 people at Johnson & Johnson in the medical device space. She was a senior FDA official in the Center for Devices and Regulatory Health. She also held a senior policy role for Senator Joe Lieberman and served on President Obama's science transition team.

BIO is the world's largest trade association representing biotechnology companies, academic institutions, state biotechnology centers, and related organizations across the United States and 30 nations around the world. BIO envisions a world where scientific innovations powered by biotechnology conquer disease, sustain our environment, and advance nutrition and health.

LEADERSHIP, SCIENCE, AND THE DESIRE TO DO GOOD

Andy Plump

As a young child, when asked what I wanted to be when I grew up, my answers had one thing in common—I was ill-suited for all of them: fireman (not tough enough), baseball player (no talent), and even veterinarian (I do love animals, but the human spirit intrigues me more).

I've always been driven by curiosity. I'd tag alongside my mother as she gardened, asking question after question. She was, and still is, a very patient woman, but I have vivid memories of her affectionately turning to me, exasperated, saying, "Andrew, it's time for school." Perhaps I was destined to become a scientist.

As with many who ultimately plod through studies in science and medicine, I was thirty-five years old when I took my first real job and it wasn't at all what I had expected.

Mother Merck

I was drawn to the pharmaceutical industry because I saw it as a place where I could apply my curiosity and love of science in a way that could potentially affect the lives of patients, not just today but for generations to come. I started in 2001 as a translational physician-scientist at the most storied of pharmaceutical companies, Merck, or for those who read Josh Boger's *The Billion-Dollar Molecule*, Mother Merck. I learned as much during my tenure at Merck as I did at any time in my life. I joined Merck at the end of an era, a few years after Roy Vagelos retired, having established a place for Merck in the annals of history.

Vagelos defined leadership and bred talent that ran layers deep. At one point in the recent past, over half of the top biopharmaceutical company R&D organizations were led by Merck alumni. But Vagelos did more than promote science and breed talent; he did good. Merck innovated and grew its business, while at the same time doing the right thing. The use of ivermectin to treat river blindness[1] and the donation of hepatitis B vaccines to China to help manage an epidemic of liver disease[2] defined the Merck "do good" mindset. So, it's no surprise that a curious boy, who failed at his attempt to play shortstop for the New York Mets (okay, who couldn't even make his high school baseball team) but became a Merck-trained translational physician-scientist, would be driven to both innovate and do good.

Leaving Merck was not easy. I had many friends and colleagues, and I maintain many close relationships that have grown during this pandemic. In between Merck and my current role at Takeda, I took a brief but highly meaningful post at Sanofi as research head. The opportunity to experience a global pharmaceutical company was an important inflection that provided me the opportunity to step into a next level of leadership in my current role as R&D head at Takeda.

Reinventing an iconic Japanese company: Takeda

They say companies are defined by their people, that a company cannot have a soul, but Takeda is different. Few companies, and, in fact, few governments standing today rival Takeda's long history. We were founded in 1781 by Chobei Takeda, an herbalist who sold traditional Japanese and Chinese medicines in Osaka, Japan. Chobei Takeda was an Omi Shonin, a group of top-tier merchants who distinguished

1. https://www.nytimes.com/1987/10/22/world/merck-offers-free-distribution-of-new-river-blindness-drug.html

2. https://www.nytimes.com/1994/06/10/business/vaccine-deal-for-merck.html

themselves through their adherence to Sanpo-Yoshi, a business principle that means "three-way satisfaction." For the Omi, the three guiding principles were good for seller (urite yoshi), good for buyer (kaite yoshi), and good for society (seken yoshi). These foundational values have defined Takeda throughout its long history. In 1940, Chobei Takeda V, a descendant of Takeda's founder, established a modernized version of the Sanpo-Yoshi that he termed the Nori. The Nori is an articulation of Takeda's values: Integrity, Fairness, Perseverance, and Honesty. Today, these values are called Takeda-ism and they continue to guide the way we work as much as they did 239 years ago.

In 2020, as we live through the worst global health crisis of our generation, it's notable that Takeda's first foray into Western medicines was 125 years ago, in response to epidemics, when we imported into Japan quinine to treat malaria and phenol for cholera. In 1914, in response to World War I, Takeda established its research division with the intent of synthesizing pharmaceuticals to offset a decline in imports from Germany.

Today, under the leadership of our current CEO, Christophe Weber, Takeda is a global, top ten biopharmaceutical company. Weber defines authentic leadership, bringing out the best of those around him. He fully embraces Takeda-ism, and has created his own rubric to articulate the Takeda values that we call P-T-R-B. Put the Patient (P) at the center of everything we do; build Trust (T) with society; enhance our Reputation (R); and the Business (B) will follow.

Weber recruited me to Takeda in 2015, providing me with an opportunity to help lead and inspire Takeda's R&D transformation. In joining this great institution under such a progressive and genuine leader, I was stepping into something much bigger. The authority granted to a position like mine, with my competitive drive, desire to win, and passion for doing good, has allowed me to play a role in addressing the COVID-19 pandemic.

The COVID-19 pandemic

With rare exception, none of us were ready for this pandemic. SARS and MERS hinted at the potential devastation that a coronavirus could have, and that animal-to-human viral transmission was a real-life scenario, not simply the makings of a sci-fi film. I am blown away by the premonition captured by a 2015 Nature Medicine paper, "A SARS-like cluster of circulating bat coronaviruses shows potential for human emergence." Not surprisingly, multiple authors on the manuscript were based in Wuhan, China. This may or may not have been a precursor to SARS-CoV-2, but we should have been more prepared. If you ask what we could have done, there is so much that it's painful to even start.

But we can't look backward, except to learn. Instead, we must seize the moment. And that's exactly what's happening.

I'll start with our efforts at Takeda. We were slightly ahead of the curve in recognizing what in hindsight is so obvious. Our vaccines head, Rajeev Venkayya, worked in George W. Bush's White House, leading a bioterrorism unit focusing on pandemic response. Venkayya was in his element when the virus hit, grasping its consequences and helping us respond accordingly. We were one of the first pharmaceutical companies to install work-from-home provisions.

By mid-March 2020, my Takeda colleagues had already begun work on a program to use plasma from people who have successfully recovered from COVID-19 to develop a concentrated antibody therapy, termed Hyperimmune Globulin (H-IG). We were also looking at our portfolio of medicines to see what might be repurposed to help the growing number of people infected with and dying from this horrendous virus. Many other companies were doing the same. Some were starting clinical trials trying to repurpose existing medicines, others were starting to discover new medicines that could be used for treatment or prophylaxis, and others were moving at lightning speed toward the development of vaccines.

Academics were publishing in open access journals about the viral sequence, evolution, structure, host interactome (the human proteins that directly interact with the CoV-2 viral proteins), the clinical course of disease, and the human immune response.

The knowledge poured in at a rate we've never seen. Nothing in the history of medicine has proceeded at a pace like this, not even close. Three features have characterized this time of COVID-19 more than any other: leadership during a crisis, the blistering pace of scientific discovery, and the goodwill of our scientific community.

Leading from the attic

Many of us work from home occasionally, but almost none of us have truly worked from home to the level necessitated by the COVID-19 pandemic. For those of us who lead large organizations, we've had to become creative in how we lead and motivate. Addressing a global organization of 5,000 employees while staring into an iPad from my attic office, with dogs intermittently barking in the background, has become routine. Agility has taken on new meaning. We're getting to know colleagues on a more intimate level.

The greatest leadership lesson I have received over my career has been the importance of creating a high-performing culture. How does one do this during a pandemic necessitating work from home? It's been easier than I could have imagined and, as always, I lean on my outstanding leadership team.

We have a strong culture at Takeda, and we work hard to make it better. We never prepared for a pandemic, but our relentless focus on people helped develop a muscle that we are now using to respond to the pandemic. The relationships and trust we have developed enable our team to be effective in a fully remote work environment. We assume good intent and help one another.

I had to assume good intent when my colleague and communications head, Colleen Beauregard, proposed a new way to connect with the global R&D organization. As the crisis was shutting down seemingly all aspects of life, I went to Charlottesville, Virginia, to bring my daughter Ali home to Boston from college. Beauregard called and suggested that I take a more casual, personal approach to engaging employees with "something like carpool karaoke." Ali and I had a great bonding time on that trip, and we shot a video while I was driving. I told everyone what this experience meant to me and how much I was thinking about them, not as Andy the head of R&D but as a fellow human navigating this uncharted territory. I've never received so much positive feedback from anything I've ever done professionally or personally. As much as I was able to help the Takeda teams, they helped me. Their personal stories poured in and were incredibly inspiring. Ali is a great guitar player, so we did end up doing a real carpool karaoke and, again, the response was incredible. We've done several more, and I throw in a little stand-up as well. I always try to have fun at work and forge real connections with people. It's more important than ever when times are tough to do just that.

We're all in this together

When thinking about biotechnology in the time of COVID-19, we will look back at this moment as a defining one for our industry. We don't have a particularly strong reputation, which is ironic given the positive mark we leave on society, and also disappointing given the well-meaning motives that drove many of my peers into our industry based solely on the potential to help people through medicine and science. What I've seen during this pandemic is what many of us have always known: our industry is filled with brilliant, humane people. This pandemic has also illuminated the fact that so many of us are driven to do good.

As we started to realize the true severity of the pandemic, several leaders across industry began having one-off conversations, exploring ways we could collaborate in our response. There was tremendous energy but perhaps a lack of focus. Rupert Vessey, an old friend and colleague from Merck, pulled a few friends and fellow R&D heads together for a phone call and we realized that many conversations with biotech companies, academics, and elsewhere were occurring in parallel. It was the Wild West. Our call evolved into a hyperfocused alliance that we now call "COVID R&D." Our goal is to add value and efficiency to a number of efforts aimed at finding treatments for this horrible disease. I feel fortunate to have staunchly committed partners and leaders driving this endeavor. From the start, Hal Barron, chief scientific officer and R&D head at GSK, said that the industry must own the pandemic response because we have the expertise, resources, and agility to do so. We have stepped up to address past global health crises, but we haven't necessarily articulated our efforts well. We have an opportunity to really demonstrate the value we bring as an industry, both in response to COVID-19 and as a critical part of society. As

Jay Bradner, president of the Novartis Institutes for BioMedical Research, said recently, "This isn't altruism. This is ruthlessly strategic." The overarching goal is for the scientific community to come up with a solution to this—and the next—pandemic.[3]

COVID R&D is a great example as one of the quickly formed collaborative efforts with good intentions, but it also has the necessary structure and rigor to make a difference. Personal agendas and priorities are being put aside and silos are being knocked down so that drug and vaccine candidates, information, data, and expertise can be shared seamlessly and in near real time. The CoVIg-19 Plasma Alliance[4] is another example where the leading plasma companies, including Takeda, are accelerating the development of potential plasma-derived therapies for COVID-19. We transitioned our own plasma-derived H-IG to the alliance in hopes that, by pooling resources, we can work more efficiently to bring treatments to patients.

The speed, agility, and willingness to adopt novel approaches to drug and vaccine development would not be possible without the trust and mutual respect that exists among leaders in our industry. Whether it's alumni from an academic institution or organization like Merck, mentors and peers, former or current investors or partners—these relationships, nurtured and tested over time, are enabling the myriad COVID-19 response efforts. Our goal is singular and agnostic to recognition, intellectual property, or direct business return—we want to solve COVID-19.

The future

COVID-19 is a deadly virus and has done damage to human lives and to the economy in ways unparalleled to those in my generation (Gen X, if you wondered). As the world approached the final days of World War II, Sir Winston Churchill rightly said, "Never let a good crisis go to waste."

---

3. https://cen.acs.org/biological-chemistry/infectious-disease/How-big-pharma-firms-quietly-collaborating-on-new-coronavirus-antivirals/98/i18

4. https://www.covig-19plasmaalliance.org/en-us#recruitment

How we work has changed forever. How we make medicines for patients is changing, and for the better. We are pushing barriers, testing conventional wisdom and the "way things have always been done." We are adopting digital technologies and sharing data in ways never imagined to this crisis. We are finding new ways to innovate, with increased speed and efficiency.

When historians look back on the 21st century, it will be remembered as the health care revolution. We have the chance to unravel human disease biology and provide solutions, if not cures, for every human disease. This is our time. We will conquer COVID-19 and we will do so with a rapidity never before seen in the history of medicine. Let's take full advantage of the learnings from this crisis and use them to propel our industry forward.

Andy Plump, MD, PhD, is president, research & development, of Takeda Pharmaceutical Company Ltd. Before Takeda, he held leadership positions at Sanofi and Merck focused on research, preclinical development, and translational sciences. Plump lives in Boston with his wife Suzanne and has three children.

Takeda is a global, R&D-driven biopharmaceutical company headquartered in Japan. Takeda focuses its R&D efforts in four core therapeutic areas—oncology, rare diseases, neuroscience, and gastroenterology—translating science into highly innovative, life-changing medicines. Its R&D organization is advancing best-in-class or first-in-class therapies to deliver new standards of care and is investing in pioneering platform technologies in cell and gene therapy, data sciences, and immuno-oncology to develop transformational, and potentially curative, treatments.

INDIA'S TIME OF RECKONING

Kiran Mazumdar-Shaw

When I boarded the flight back to India from New York on March 7, the city had just identified one COVID-19 case and the World Health Organization (WHO) was still deliberating on the severity of the contagion threat to the world. When I arrived in Bangalore on March 9, WHO finally declared COVID-19 as a global pandemic, but India had already swung into action. I was temperature profiled at the airport and required to fill in a quarantine form that asked for information on the flight, seat number, transit airports, cities visited over the past two weeks, and of course contact information. I was skeptical of its utility but was super impressed when I got a call from a government health official asking how I was feeling and who said they would call me every day for two weeks to make sure that I did not develop any flu-like symptoms. By March 15, India had started quarantining all international travelers in dedicated quarantine centers, banning all international flights, testing quarantined passengers, and shifting those who tested positive into hospitals. Fast forward to March 24, and India decided it was time for a twenty-one day complete lockdown. India's infected population was just over 500 compared to Italy, which had more than 9,000 cases when it locked down, or the UK, which shut down when figures touched 6,700.

Rapid response and preparedness planning

This early and rapid response to a deadly contagion helped the world's second-most-populous nation to equip itself to deal with the pandemic. The lockdown was utilized for a massive preparedness planning exercise to assess the biomedical, socioeconomic, and demands of public health infrastructure to withstand the onslaught of an invisible enemy. Multiple task forces were set up to pursue efforts in mission mode. It was evident that testing, tracking, quarantine, and treatment was the right approach to stop the spread.

For a country of 1.3 billion people, testing even at a low level of 1,000 per million would require 1.3 million tests. By mid-March, thirty government labs were testing at a combined rate of only 10,000 tests per day. At this point we moved the government to expand capacity through the induction of private diagnostic labs, which could quickly ramp up testing to 100,000 tests per day. The government welcomed this and enrolled a number of private labs, but it soon became apparent that test kits were in short supply. India was completely dependent on imported RT-PCR kits, but surging COVID cases in the US meant a rationing of test kits for the rest of the world. A quick call for action from a number of small diagnostics kit manufacturers saw three indigenous test kits being assembled in a matter of weeks. By the first week of April, we had created an indigenous testing capacity of 150,000 tests per day.

Repurposing: A new mantra

With just 0.55 government hospital beds per 1,000 population, India faced a huge capacity challenge. With some of the world's best health systems buckling under the strain of tens of thousands of coronavirus patients, the prognosis for India was grim. However, a rallying call to the private sector saw an innovative model being created by sequestering hotels to quarantine mild and asymptomatic COVID-19 patients, who could be shifted to nearby hospitals if and when their condition worsened. With tourism impacted, hotels and hotel staff were being gainfully harnessed to repurpose their businesses and protect jobs. The government, on its part, created a capacity of 100,000 hospital beds in state-run hospitals and large indoor stadiums. The pan-India exercise bolstered the country's COVID-critical hospital infrastructure by adding nearly 400,000 isolation beds by the end of April.

India entered the COVID-19 pandemic with an installed base of ventilators that was just short of 45,000 units, with the majority concentrated in private tertiary care hospitals. Going by the experience of countries like Italy and the US, it was becoming evident that India would need to ramp up ventilator manufacturing capacity, and quickly. Some of India's biggest automobile makers collaborated with small ventilator makers, using their engineering skills to repurpose production lines for ramping up much-needed ventilator manufacturing capacity in the country. From a mere 5,500 units a month in March, India plans to increase output to 50,000 ventilators a month by the end of May.

Likewise, India's large garment industry repurposed itself for mass production of PPE—coveralls, masks, gloves, and other protective gear.

Indian alcohol distillers repurposed themselves to produce hand sanitizers, which were in short supply in the country when the outbreak hit.

The pandemic was a moment of awakening for India because we demonstrated that we could very quickly repurpose many of our industries for the war effort against COVID-19. Today India's stockpile of medical supplies is in a state of satisfactory preparedness.

A scientific awakening

Serendipitously, the COVID-19 crisis has led to the discovery of a deep reservoir of scientific and engineering competencies within India. The emergency has galvanized the Indian scientific community to collaborate like never before in finding interdisciplinary solutions to give the country an edge in this high-stakes battle against a deadly virus.

Protecting a billion-plus population from a pandemic in a country that traditionally spends insufficiently on science was never going to be easy. Yet Indian science proved itself resilient, resourceful, and proactive.

Scientists in India worked overtime to come up with innovative solutions to both big and small problems across a range of biomedical fields.

From mathematical models to predict viral spread and containment to apps for contact tracing, from RT-PCR and serological test kits to treatments and vaccines, there was a concerted effort to collaborate on a common platform of resource and knowledge sharing like never before.

In the past six weeks, India has done an exemplary job of creating consortiums for pooling complementary skillsets. Task forces have been set up to hunt for promising leads for COVID-19 vaccines and drugs. R&D labs, academic institutions, start-ups, and small enterprises have been encouraged to come forward with ideas, find industry partners, and scale up quickly. Everyone has worked for the collective good in this hour of crisis.

India's role in vaccines and therapeutics

India's vaccine producers, who are among the largest global suppliers of vaccines for polio, meningitis, pneumonia, rotavirus, BCG, measles, mumps, and rubella, are now focused on independently developing or partnering with innovators around the world to produce vaccines against the SARS-CoV-2 virus.

Serum Institute of India, one of the world's largest vaccine makers by volume, is collaborating to mass-produce a vaccine being developed by the University of Oxford.

Biocon Biologics has initiated a clinical trial in India to study its anti-CD6 monoclonal antibody, Itolizumab, for the treatment of severe COVID-19 patients with complications arising from cytokine release syndrome. Research efforts are also afoot to develop therapeutic vaccines based on neutralizing antibodies against COVID-19.

Conclusion: We are all in it together

John Kenneth Galbraith, American economist and diplomat, had famously characterized India as a "functioning anarchy." The COVID-19 crisis has shown that, when needed, India can come together to rise above the anarchy. It has taken a deadly viral pandemic to uncover the nation's resilience, resourcefulness, and innovativeness.

India's strategy is in stark contrast to what we are seeing in the US, which seems to be directed toward winning the race for the world's first billion-dollar antidote. Except for a handful of initiatives in the Big Pharma world, we are witnessing knowledge competition rather than knowledge collaboration. Instead of joining forces, companies are trying to outsmart each other. The focus is on beating the competitor, not the virus. The motivation is profits, not patient lives.

At a time when humanity is facing one of the biggest threats to its existence, "success" should not be about making billions of dollars, but about protecting billions of lives.

The global scientific community needs to focus on ensuring affordable access, whether it is to a new vaccine, a new drug, diagnostic kits, or protective masks.

Clearly, it is time to create a new paradigm to address the escalating crisis by ensuring fair, transparent, and equitable access to essential medical supplies and any future vaccines developed to fight COVID-19. Ultimately, the greatest lesson that COVID-19 can teach humanity is that we are all in this together.

Kiran Mazumdar-Shaw, chair of Biocon, is a pioneering biotech entrepreneur, a health care visionary, a global influencer, and a passionate philanthropist. The impact she has made as a leading woman in science has made her a role model in India. She is committed to equity in health care through access and affordability as she pursues a path of making a difference to billions of lives globally.

The Biocon Group is a globally recognized biopharmaceuticals and research services enterprise driven by its passion to enable affordable access to essential drugs for patients worldwide. Biocon is driven by innovative science and research partnerships to develop biosimilars, novel biologics, and complex generics for cancer, diabetes, cardiovascular, and autoimmune diseases for patients and health care systems globally.

CONNECTING THEORY, DATA, AND POLICY FOR A COMMON CAUSE

Karen Bernstein

In the information business, what you're doing at 9 a.m. on Monday isn't necessarily what you will be doing at 9:15. BioCentury was founded in 1992 to provide mission-critical analysis and data to biotech companies, their investors, and regulators. Over the past twenty-eight years, we've been through the HillaryCare crisis, the collapse of the genomics bubble, and the 2008 financial crisis, not to mention numerous smaller crises in between, so we know how to pivot. Thus when it became clear that COVID-19 was an all-hands-on-deck moment, we pivoted everything and everyone at the company to focus both on what we needed to do to fulfill our public mission, and what we needed to do to make sure we could survive and navigate all the (known) unknowns as well as the (unknown) unknowns that would surely surface.

Our external job, as we saw it, was to provide as much information as rapidly as possible to as many people as possible about the virus and its biology, as well as all the countermeasures being developed globally by both companies and institutions: vaccines, therapeutics, and diagnostics; human clinical trials and preclinical animal trials, and in vitro work. Everything. Theories and data. Regulatory policy and public policy.

Throughout, we have aimed to act as a central platform for connecting people in the industry while they are under lockdown: for example, with surveys to inform industry scientists and executives of what their peers are thinking, as well as webinars and podcasts to provide different formats for airing discussion and ideas.

It was an easy decision for us to put all the drug science in front of our paywall and make it free to the world.

As the volume of our data and analysis increased, we concluded we needed a dedicated portal to make it easy for companies and researchers not only to find out what others were doing, but to make it simple to contribute their ideas, compounds, and news of their clinical trials to others working on the pandemic.

From idea to execution, it took us nine days to build a portal, including designing the user interface, implementing new technology, and collecting and curating information from multiple sources to populate it. Four groups inside the company dropped what they were doing to turn this around in a fast and coordinated way. For me, this was a powerful reminder of the incredible resourcefulness, ingenuity, and ability of the private sector to come up with solutions and do what is needed in times of crisis. No government or NGO could do this faster or more efficiently.

In parallel, in response to the proliferation of small, uncontrolled clinical trials being started all over the world, a group of colleagues at several venture capital firms and pharmaceutical companies had started talking about how to collaborate to bring forward the best therapies and vaccines as quickly as possible.

One of their (justified) fears was that the studies being run at random would waste resources because the compounds being tested were not the best compounds, and the trials would be too small to yield meaningful results. Even more worrisome, patients in these studies wouldn't be available to test the best compounds in well-controlled studies that would yield actionable data.

In less than a month, this group quickly coalesced into a collaboration named COVID R&D, which launched with fifteen biopharma companies and two VCs as its core. Its mission is to share data, identify the molecules—from among the hundreds being proposed—with the strongest rationale for advancement into human studies, and put those molecules into well-designed trials.

Because of the work BioCentury had been doing to catalog everything in development, or with a potential rationale for development, COVID R&D came to us to serve as a central source for investigational products they could draw from. BioCentury built out the portal to include 407 molecules and counting, with details such as target, molecule type, and references to help the pharma group streamline their evaluations.

The result has been a collaboration in which BioCentury organizes and channels publicly available information to the consortium, while serving as a resource for the broader biomedical community.

In parallel, BioCentury had set up an internal working group to continue adding value to the portal. The next step was to create a gateway that enables researchers and companies to submit compounds to be included in our COVID-19 Resource Center.

Five days later, we expanded the gateway to enable companies and researchers to submit details of investigational products directly to any of the seven working groups that are part of COVID R&D: repurposing of clinical phase compounds, preventive vaccines, novel antibodies, novel small molecule antivirals, repurposing of preclinical stage compounds, data sharing, and clinical trial acceleration.

Meanwhile, BioCentury has continued to write about the importance of the master protocols being developed for testing of multiple compounds in a single study, or for testing multiple compounds in different trials but using the same endpoints and control group standards.

BioCentury has been championing master protocols for more than two decades. But while this approach is now used in some cancer trials, it hasn't seen widespread use. Maybe COVID-19 will change that. These kinds of trials can be incredibly efficient in terms of time and money, weeding out the drugs that will not make it while the best ones can advance more rapidly.

The biopharmaceutical world has embraced this model during the pandemic; the question is whether the lessons of COVID-19 will lead to broader acceptance of master protocols in other disease areas, or whether we will go back to the status quo ante when the crisis is over. That would be a huge missed opportunity to speed up the development of lifesaving treatments for other diseases.

We have also been thinking about the regulatory systems around the world and how they have adapted to the pandemic. They have loosened up regulations in many ways, seeking to balance the life-and-death urgency of the crisis while not eroding safety.

The push-and-pull between speed and safety is not new. But today's pandemic is a giant experiment, and we will see after the crisis what lessons regulatory bodies can take forward to get new treatments to patients faster.

This holds not just for companies developing drugs, vaccines, and diagnostics for COVID-19, but also for all the drugs being developed for other diseases whose trials have been disrupted by the crisis.

Meanwhile, even as the crisis goes on, so does the business of innovation. We have been providing insights about how biotech companies can survive while the financial markets and drug development partnerships are in semi-shutdown.

We are also thinking about what life in the biopharmaceutical world will be like after the pandemic ends, or at least is under control. How soon will things return to normal? Will there be a new normal that's different? We don't know, but BioCentury's role has always been to think and write about the big picture to help CEOs and other senior managers who are fully employed in their day jobs.

Some big positives could come out of this crisis.

One is for the public and politicians to see that the biopharmaceutical industry isn't an evil and heartless corporate machine, despite the narrative promoted by mainstream media and populists from both sides of the political spectrum.

These companies are doing important work, in the most profound sense of that phrase. Yet while I see the National Institutes of Health and the Bill and Melinda Gates Foundation and the World Health Organization in the press daily, I see no mention of who is really doing the heavy lifting: the biotech and pharmaceutical companies that are the only ones equipped to discover, develop, and manufacture the hoped-for diagnostics, therapeutics, and vaccines.

Especially vaccines. Creating vaccines is hard and the coronavirus family has been challenging. So is large-scale manufacturing for any vaccine. I believe we will get there, probably in stages where the first therapeutics and vaccines aren't as good as the ones that will come later. But no one should forget that the capabilities to develop these lie in the private sector, particularly for manufacturing vaccines at scale. Academic scientists know how to discover new biological insights, but they don't know how to develop drugs—it's just not their job.

And industry is doing all this for free. As in the Ebola crisis and other pandemics, biopharma companies have banded together to do what must be done, with no fanfare and no publicity. They should get full credit for their efforts. These companies are hardly perfect—just like the rest of us—but politicians who find them an easy target should open their minds and put down their cudgels. It is equally the job of reporters at mass-market business and consumer media to open their minds and understand and report on how the world really works, and who will ultimately provide the solutions to today's public health crisis.

Last, but surely not least, if any good comes out of this crisis, we hope it will be that governments and payers will finally, finally accept what many in the industry have been saying for years: that the market for antivirals, antibiotics, and diagnostics is broken. That if we want to be properly prepared for the next crises—and many of these, especially for drug-resistant bacterial diseases are (known) knowns—we must be willing to pay for the diagnostics and treatments that will be required.

Being willing to pay $50,000 for a cancer drug that provides two months of extended life while balking at paying $700 for a lifesaving antibiotic is insanity. Ditto for modern molecular diagnostics. Yet that is where we are at today. Some diseases, like COVID-19, can't be predicted. But others can, and we should pay companies to develop and stockpile drugs for them. Who knows how many deaths we can prevent the next time around.

Karen Bernstein, Ph.D., cofounded, co-owns, and is chair of BioCentury. She is a director at Ovid Therapeutics Inc. and at Codiak BioSciences Inc., is a trustee of the Keck Graduate Institute, and is a member of the board of advisers of KGI's School of Pharmacy. She serves on the board of overseers of Scripps Research.

Since 1993, **BioCentury Inc.** has been internationally recognized as the leading provider of value-added information, analysis, and data for biopharmaceutical companies, investors, academia, and government on the strategic issues essential to the formation, development, and sustainability of life science ventures. BioCentury employs a fully integrated multimedia platform—including publications, video, online data solutions, and conferences—to provide authoritative intelligence about corporate strategy, partnering, emerging technology, clinical data, public policy, and the financial markets.

LOVE IN THE TIME OF CORONA

Ron Cohen

In Gabriel García Márquez's classic book, *Love in the Time of Cholera*, the protagonists Florentino and Fermina have a tumultuous affair in their teens, then reunite in older age, having been apart for fifty-one years, nine months, and four days. They make love on a river cruise. As the ship docks at its final port, Fermina sees people on shore she knows, and fears that she and Florentino will be disgraced. Florentino tells the captain to raise the yellow flag, signifying cholera, so that no one will board. But no port will then accept them, and they are condemned to wander the river forever.

In a time of plague, what must we do to navigate the river, uncertain if, when, or how we will again arrive at a safe harbor? For business leaders, the challenges are compounded: we must contend not only with personal and family upheavals as others do, but also with the impact of such upheavals on the people on whom our businesses rely: our associates, customers, partners, and more.

This is what I have learned so far:

Love of one's self, family, and friends

"*At a cardiac arrest, the first procedure is to take your own pulse.*"[1] Notwithstanding popular conjectures to the contrary, business leaders are

1. A key "law of the hospital," in Samuel Shem's "*The House of God.*"

human. Before addressing the needs of those for whom we are responsible, it's critical that we first address our own. In my case, I have looked for constants to serve as anchors through the crisis—for example, reserving time each day for my nuclear family, and staying in touch with extended family and friends by text, email, and Zoom. (I believe the pleasures of Zoom cocktail parties are likely to endure past the crisis.) I have also kept to my usual diet and have bought exercise equipment for use at home to substitute for my former almost-daily gym routines.

In addition, I reserve time each day to catch up on news concerning the pandemic. As a physician, I am particularly interested in the latest clinical studies, disease trends, testing capabilities, policies, and impacts on our industry and society at large. In this way I can evolve my views of where the river is taking us, and of how much longer we are likely to be confined to the ship. I can then help communicate important information to Acorda's associates and others in my sphere.

Love of one's work, purpose, and teammates

In times of upheaval, just as for individuals in their personal lives, a company's associates need compasses to help guide them through the storms. Two of the most valuable such compasses are the company's mission and its culture. The mission defines the "what" of the work, and ideally is clearly defined and understood by all its associates. At my company, Acorda, our mission is to develop therapies that restore function and improve the lives of people with disorders of the nervous system. Culture is harder to define—as with the concept of "beauty," you know it when you see it. A company's culture guides the "how," and even more importantly, the "why" of the work its people do.

Acorda's culture has been built on a set of Principles and Values (P&Vs)[2]. We incorporate them into our day-to-day conversations, we celebrate and give awards to those who exemplify them, and we provide them on desk ornaments for all associates; they even are etched two stories high onto the glass walls of our headquarters building in Ardsley,

2. http://www.acorda.com/careers/working-at-acorda

New York. They are expressed colloquially, some tongue in cheek; we don't want to take ourselves too seriously, but we do take the culture, and the work we do to improve patients' lives, deeply seriously. One more thing: other than on the website, there's an "inside" phrase that always appears with the P&Vs wherever they appear: "Fresh Fish Sold Here." There's a story that goes with it, which has become part of the company's lore. What's important is what it signifies: that any list of principles, values, or the like is just a string of words; they are given life and meaning only by the actions of those who subscribe to them.

During the pandemic, we've found that our P&Vs have served as a lodestone. For example, **"Communication, Communication, Communication"** and **"We tell it like it is"** have guided our leadership team members to send frequent emails to the entire company, including at our headquarters and manufacturing facilities, and in the field. We frankly communicate news and prospects related to the pandemic, as well as about the company's business, challenges, and progress. I also periodically address the company by video and field questions. Our P&Vs of **"We will find a way or make one"** and **"Teamwork . . . uh, huh!"** have underpinned the remarkably rapid adaptation of our teams to working from home, to finding new, effective ways of communicating and working with each other and with the health care professionals and patients we serve. They continually devise new programs to advance Acorda's work and mission, as expressed in our P&V **"Therapies or bust!"**

The stresses of COVID-19 have been superimposed on what was already one of the more challenging periods in Acorda's twenty-five-year history; since a court overturned the patents on our lead product over eighteen months ago, we've been forced to implement painful reductions that cut our workforce in half and put our development programs on hold. Despite these enormous stresses, in April 2020 we experienced a remarkable, and moving, testament to the impact of our company's culture, when Acorda was named to Fortune's Top 10 list of the Best BioPharma Companies to Work For. An overwhelming 94 percent of Acorda's surveyed employees agreed, versus an average of 59 percent for the industry.

Love in the time of corona

> "The world is too much with us; late and soon, Getting and spending, we lay waste our powers. . . . We have given our hearts away, a sordid boon!"—WILLIAM WORDSWORTH.

We do not know if or when the ship will return to safe harbor; we can only project. COVID-19 is a crisis. It is also an opportunity to reimagine our priorities and responsibilities—to our businesses, our society, our fellow human beings, and our planet itself. It's an opportunity to recalibrate what will make our time on earth most meaningful. In the late second century AD, an estimated 10 percent of seventy-five million people in the Roman Empire died of what may have been smallpox, as the plague spurred the government and citizens to come together, adapt, and rebuild a society that emerged stronger than ever.[3]

This is our time. This is our opportunity.

Ron Cohen, MD, has been a leader in the biotechnology industry for over three decades. He is president, CEO, and founder of Acorda. Cohen served as chair of the board of the Biotechnology Innovation Organization and continues as a board member. He has received numerous awards, including the NY CEO Lifetime Achievement Award and the Ernst & Young Entrepreneur of the Year Award; he is an inductee of the National Spinal Cord Injury Association's Spinal Cord Injury Hall of Fame; and he was awarded Columbia University's Alumni Medal for Distinguished Service. He was recognized by PharmaVOICE Magazine as one of the 100 Most Inspirational People in the Biopharmaceutical Industry. Dr. Cohen received his BA from Princeton and his MD from the Columbia College of Physicians and Surgeons. He is board certified in internal medicine.

---

3. Edward Watts, "Why Romans Grew Nostalgic for the Deadly Plague of 165 A.D.," *Zocalo Public Square*, April 26, 2020, https://www.zocalopublicsquare.org/2020/04/26/roman-empire-smallpox-plague-lessons-covid-19/ideas/essay

Acorda Therapeutics Inc. develops therapies that restore function to people with neurological disorders. The company has 320 associates and has brought three therapies to market, including Zanaflex Capsules for the treatment of spasticity, AMPYRA® to improve walking in patients with multiple sclerosis, and INBRIJA™, an inhalable powder form of levodopa for adults with Parkinson's disease. Acorda has been named among the Fortune 100 Best Medium Workplaces in the US, the Fortune 100 Best Workplaces for Women, for Millennials, and for Baby Boomers, and is among Fortune's Top 10 Best BioPharma Companies to Work For. For seven years, Acorda has been recognized as one of the Best Companies to Work for in New York State.

FROM DIFFERENT PARTS OF THE WORLD, WE CAN FIGHT COVID-19

Samantha Du

During this unprecedented crisis of human suffering in the age of COVID-19, I, like many others in the biotech community, have tried to reflect on how our industry, at the front lines of the response effort, can better evolve and grow to tackle the human health challenges of today and tomorrow. Though we all know that our industry will play a critical role in improving human health and elevating the state of care for those facing disease, whether rare, chronic, or otherwise, the magnitude of this crisis serves as a catalyst for all of us to reflect on how we, as an industry, can learn from this shared experience and best use our collective capabilities to shape the future.

As an emerging global biotech company with major operations in China, Zai Lab witnessed the impact of the first big wave of COVID-19. Our team took decisive steps to put the interest of our patients and our employees first. A key goal was to minimize the disruption to patients enrolled in our clinical trials, many of whom are seriously ill cancer patients. We quickly developed new procedures for patient evaluation and revised our drug distribution protocols to ensure safety and study data integrity. We also communicated quickly with our investigators and trial sites to monitor and ensure compliance. Through these swift actions we were able to mitigate against significant patient drop-off relating to the COVID-19 outbreak in our clinical trials, and largely avoided the delay of study completion. On the commercial side, we were fortunate to launch Zejula, a lifesaving medicine for ovarian cancer patients, in January, shortly before the outbreak in China. Our commercial team rapidly deployed a creative online platform for continued physician education and, importantly, to enable patients to receive timely access to our drug. Our team demonstrated its ability to quickly adapt to address the key challenges posed by the pandemic.

We also took decisive action to help frontline health care workers and our biotech colleagues around the world, while preparing and implementing contingency plans to minimize disruption to Zai Lab's operations. Initially, we donated cash and medical supplies to hospitals in Wuhan. More recently, with the proliferation of the virus in the US, Israel, and other parts of the world, we procured and donated masks and PPE to biotech companies to support their essential lab operations. We also shared operating protocols and lessons learned through our own experiences with our partners outside China. We received overwhelmingly positive feedback from our colleagues and peers at other biotechs on the value and substance of our collaborative knowledge sharing activities.

In this crisis, members of the biotech industry around the world demonstrated that we are one global family. We have also demonstrated that what we do in China can help the global health care community.

This makes me reflect on my own journey over the past twenty years. When I first returned to China in the early 2000s, having spent seven years at Pfizer central research in the United States, I envisioned China over the next few decades taking an increasingly pivotal role in advancing the global biotech industry. I also recognized that, at my time of return, innovative biotech companies were nonexistent in China. Fast forward twenty years, and the progress made and capabilities built by the biotech industry in China are virtually unprecedented. This was on display during the response to the COVID-19 outbreak. The virus sequence was published in days, and many molecular and antibody diagnostic tests were developed within weeks. More than twenty local companies in China began new drug or vaccine development programs against COVID-19, with several programs already in clinical trials. Although the growth has been impressive, I foresee that China's major contributions to our common mission to improve global human health remains ahead of us.

My vision for China's role within the global health care arena is embodied in Zai Lab. My first key strategic vision when founding the company six years ago was to quickly embrace China's regulatory reform and changes to the reimbursement landscape, to bring sorely needed, innovative drug products to Chinese patients. At the time, over 70 percent of branded, FDA-approved drugs were not available to patients in China. The regulatory reforms initiated years ago have resulted in the approval in China of a multitude of internationally developed drugs over the past several years. Zai Lab pioneered the business model of partnering with our US and European biotech peers, licensing their late-stage clinical assets, and developing and commercializing these products in greater China as a way to get these products to patients in China faster. As an example, in December 2019, our product Zejula became the first in-licensed innovative medicine by a Chinese biotech originally developed by a US biotech company to be approved by the drug regulatory authorities in China.

My second, longer-term, goal is to contribute China's capabilities and successfully integrate them into the global biotech value chain. Just like the important biotech communities in Europe and Japan, I am confident China will become a fundamental part of the conversation in global drug development; it goes without saying that Zai Lab will be a major contributor to the global scientific and medical communities.

China has long been part of the global manufacturing supply chain, but what I see as a more defining moment for China is when it is playing a more central role in advancing novel translation medicine approaches and methodologies and becoming a critical participant in global clinical development pathways. Given the recent addition of

China to ICH, the improved quality of Chinese clinical data, the large patient population, and still relatively large naive patient population, development plans that contemplate China can significantly accelerate overall timelines and enable therapies to quickly reach patients who need them the most. We have seen several examples of this, whether it be in the form of the participation by Zai Lab and Ju or other Chinese biotech companies in multiregional clinical trials with international partners, or internal drug development initiatives in China by global pharmaceutical companies. China is increasingly being recognized as an indispensable part of any development program, with the additional benefit of enabling drug approval in both the US and China, two of the largest markets in the world. A key reason is that the quality of innovation within China has substantially improved. In fact, innovation in China has become paramount for many, both in the domestic biotech industry as well as the China R&D operations of multinational pharma and biotechs. Case in point: the first company I established after returning to China was the first to secure regulatory approval for a China-innovated and domestically made drug for a major cancer type. Several other China-innovated drugs have recently made it to the world stage, and we're just at the tip of the iceberg.

We are seeing the rising role of China in global drug development during the COVID-19 pandemic. As China is slowly returning to a post-COVID-19 "normalcy" before its Western counterparts, China is resuming patient enrollment more quickly and picking up the gap left by other regions, many of which remain heavily affected by the pandemic. China's current pivot to "normalcy" serves as an important reminder that the market provides a hedge to our industry's all-important goal of timely testing of experimental medicines in patients with unmet medical needs. Additionally, China's contributions to numerous multiregional clinical trials for potential COVID-19 treatments as well as its scientific understanding of this new invisible enemy will benefit the industry and the entire global community.

This recent pandemic has illuminated a glaring vulnerability of our world to the threat of disease, and I envision China stepping up in an even more significant manner in the coming years to play major in innovation, discovery, and development not just for China, but for the world in an globally coordinated efforts to tackle the challenges of tomorrow's "new normal." This can be substantiated by the increasing amount of global capital invested, heavily in innovative health care, in both private and public sectors, more maturing local scientific and entrepreneurial talent, improving regulatory environment, opening both mainland and Hong Kong stock exchange boards to biopharma companies, more favorable government policies, and the rapidly growing aging population. Cancer has no borders. COVID-19 has no borders. Thus, science and innovation cannot have borders. As we often say, we are all in this together.

Dr. Samantha Du is a Chinese life science industry pioneer, global drug developer, passionate serial entrepreneur, and investor. She is the founder, chair, and CEO of Zai Lab, the founder of Quan Capital, and a co-founder of Chi-Med. Seven of the life science companies she either founded or was an early investor in are publicly traded, each with multibillion-dollar market caps. Du is widely recognized for her work on several groundbreaking health care policy reforms in China promoting innovation and regulatory harmonization with global practices.

WHAT I LEARNED FROM HAVING COVID-19

Cedric François

On March 14, 2020, at 10 a.m., I got the call. I had tested positive for SARS-CoV-2, the virus that causes COVID-19. At first, I was incredulous, trying to understand where I may have contracted the pathogen. Next, I got worried. Who had I been in contact with? Who else may I have transmitted the virus to? Finally, a nagging feeling in my stomach that maybe, at forty-seven years of age and male, I may be at risk of a bad outcome.

In the midst of these emotions, I also had my company to think about. Apellis, the gift that Pascal Deschatelets, Alec Machiels, and I were given when we embarked on our entrepreneurial adventure in 2009, and which now had more than 275 employees and two programs in Phase 3 clinical testing. I became gravely concerned about what a true global pandemic would mean for our growing company. I also worried for our industry, as the complexities and hardships caused by the pandemic would not be specific to us.

Our employees were worried, our stock as well as the entire stock market were in freefall, and I was losing sleep over our clinical trials. In dozens of countries, travel bans were being imposed that could make it difficult to get our investigational therapy to patients in our clinical trials who so desperately needed it.

Top of mind were our OAKS and DERBY trials, with target enrollments of 600 patients each and the potential to create the first treatment for geographic atrophy (GA), the advanced form of macular degeneration, a blinding disease that affects one million patients in the United States alone. GA affects the category of patients so much at risk when exposed to COVID-19: the elderly. How were these patients going to get to the clinic to complete our enrollment, which was oh-so-close to the finish line? And more importantly, what could we do to make sure the patients who were already enrolled could safely receive their treatments, which require a monthly office visit to a retinal practice?

As these and many other questions swirled through my head, I spoke with our board of directors, and we decided that radical transparency was needed. I created a video to share my diagnosis with our employees and also published a post on LinkedIn. The positive response was overwhelming, filled with understanding and empathy. Most important, the support and zeal coming from the Apellis employees touched me deeply.

It was time for action, but I was sick. Not too bad, initially. Good enough for phone calls and emails. But then, after seven days, when I thought I was getting better, I spoke with a dear friend and physician in the Netherlands, who warned me of the "second phase." And that evening, it struck. For four days, I became sicker than I had ever been. I was hurting terribly in every part of my body. I completely lost my sense of taste and smell. And then, as suddenly as it came, it was over. I was still tired for another ten days, but on the mend.

Overall, I was fortunate—the disease affected me much less than so many others, and Apellis held its own remarkably well. Our employees remained safe and resilient, and did an extraordinary job of ensuring the safety of our patients in our clinical trials. As I reminisce on my experience with COVID-19 as both a patient and CEO, a few observations are branded in my mind:

COVID-19 brought out the better part of humanity

And it has done so in spectacular fashion. The pictures of health care workers on the front lines, putting their lives and those of their families at risk, are as gut-wrenching as they are inspiring. The images of kids distributing groceries to vulnerable people and special initiatives emerging everywhere remind us that we can all make a difference. It is also a reminder of how important it is for all of us in the biotech industry to bring compassion to our work and to the decisions we make every day.

The biotech world is a place of wonder

Our world is populated by the best and the brightest, men and women who study for decades and work night and day to bring new medicines to people who need them. In the past few decades, our industry has too often been in the news for the wrong reasons and—based on a few bad actors—has been cast as a greedy industry that profits at the expense of people's health. This poor image belies the dedication of the overwhelming majority of those who work to discover new medicines. Never was this more apparent than now, when countless companies engaged in record time to test hundreds of possible approaches to contain the pandemic. But innovation across biotech was only made possible by the tireless dedication of the men and women at the US Food and Drug Administration (FDA) and other regulatory organizations, who have been working urgently and nonstop, with no end in sight. Out of the spotlight, they have been heroes in helping to advance therapies for COVID-19. The resilience of our entire industry and its dedication to help humankind is something that I am proud of, grateful for, and of which I like to think we at Apellis are proud stewards.

Increased collaboration for the greater good

Throughout this pandemic, we have seen shining examples of the government and industry working together to advance potential vaccines and therapies for COVID-19. The FDA has reviewed COVID-related programs in record time, setting a precedent for future accelerated drug development. Companies have come together, concerned about patient safety and progress. In retinal disease, we were thrilled to work with the heads of medical and development from several other companies to promote guidelines on safely conducting clinical trials during the pandemic. On a personal level, I had the great fortune of being part of a collaboration between the Commonwealth of Kentucky and local hospitals and universities to measure how much immunity is generated by COVID exposure, and to identify those with the best immune response as donors for high-quality plasma treatments. These organizations, some of which were traditionally competitors, came together in weeks to address a critical need. The dedication and collaboration have been inspiring and give me great hope for our future.

What will the post-COVID world look like?

Initially, it will be hard. I am heartbroken by the tens of millions of people who were forced to file for unemployment. We have a natural tendency to believe that this will be temporary and that we can go back to the way things were before as soon as the virus is under control. I am doubtful this will be the case. The nature of many businesses is likely to change forever.

In the midst of this bleak prospectus, however, it is important to focus on the positive. And there are many things to be excited about. For our industry, I would hope that we can be a shining example of how science can make the world a better place; how, when invisible enemies like SARS-CoV-2 ambush humanity, biotech is the first line of defense with the brightest and most dedicated people leading the charge.

For our world in general, I hope that we wake up to our vulnerabilities, not as a race, but as harbingers of something far greater than ourselves. If we were given anything by SARS-CoV-2, it was time—time that has afforded us an opportunity to be home with our families, time to appreciate our beautiful planet, and time to rediscover our appreciation for others and for what we have.

What I wish from the bottom of my heart for humanity and our planet is that we come out of this crisis more aware, more present, and more grateful for the abundance that we are given in this life. Abundance in love, in life, and in kindness. Because for a moment in time, in the spring of 2020, we showed our best as humans toward one another and toward the immense need of those less fortunate among us. As a society, as a planet, we have a lot of pain to go through still, and the path to healing will be long. But through our journey to recovery,

I hope and I am confident that we will remember the good lessons of the pandemic of 2020.

Cedric François is a Belgian physician-scientist and cofounder and CEO of Apellis, a global biopharmaceutical company pioneering therapies for conditions caused by the dysregulation of complement. Complement is one of the oldest parts of our immune system and believed to be at the root of many human diseases.

Apellis leverages courageous science, creativity, and compassion to deliver life-changing medicines. We aim to develop best-in-class and first-in-class therapies for a broad range of diseases that are driven by uncontrolled or excessive activation of the complement cascade.

THE VIEW FROM INSIDE COMPANIES

A DEFINING MOMENT

George Scangos

I chose to spend my life in the biotechnology industry trying to bring medicines to people who need them. The reasons have never been clearer. We are now on the front lines fighting a virus that has killed so many people already, with the potential to kill millions more, and has shut down society and crippled economies around the world. Absent drugs and vaccines, it will take years for sufficient herd immunity to develop so that society can resume. In that case, we would be forced to compromise between continued isolation and the consequent social, emotional, and economic devastation, or relaxing our isolation at the expense of increasing deaths. It is not being overly dramatic to say that the nature of our society over the next decade will depend on our success and how quickly we can achieve it.

VIR was founded to take on major global infectious diseases, including potential pandemics. We did not expect to be confronted with such a serious pandemic so soon, but here we are. For the employees at Vir, many of whom have devoted their lives to preventing and treating infectious diseases,this is our moment: *this is why we exist.* Since early January, we have been working long hours, seven days a week. We have prioritized COVID-19 activities above all others. We have accelerated timelines, worked with current and potential collaborators, interacted with regulators, and condensed a process that normally takes eighteen months into four months. We plan to begin clinical trials this summer and to complete them in the fall. If successful, we have planned for a drug supply of millions of doses by the end of the year.

We are taking multiple approaches to bring forward potential therapies for prevention and treatment of COVID. Our first clinical programs are likely to be antibodies. Our scientists have repeatedly demonstrated the ability to isolate therapeutically relevant antibodies from patients who have recovered from infections. Two examples are mAb114, which is one of the two treatments that showed some efficacy in the recent Ebola trial conducted in the Congo by NIH, and a pan-influenza-A antibody that recognizes every strain of Flu A that has arisen since the 1918 pandemic and will start Phase 2 trials for flu prophylaxis later this year. A number of years ago, we had isolated antibodies from survivors of the SARS-1 epidemic. When SARS-CoV-2 emerged, we went back to those SARS-1 antibodies to assess whether any of them were capable of cross-reacting with SARS-CoV-2. And of course we also screened antibodies from patients who recovered from CoV-2 infection.

We believe that antibodies that cross-react with an entire viral family, and therefore recognize an epitope that is highly conserved, are likely to have advantages as therapeutic agents because resistant mutants are less likely to arise. Cross-reacting antibodies that also are potently neutralizing do exist—but they are rare and it takes skill to find them. Our lead antibodies are examples and we are moving them rapidly toward the clinic while simultaneously characterizing additional antibodies that could pair with them. Preliminary data as of this writing indicate that we have additional antibodies that act synergistically with our lead and could reduce the dose, and thus increase the number of treatable patients manyfold.

Our second approach comes from our siRNA collaboration with Alnylam. We have been moving forward an siRNA targeting hepatitis B, which has generated good data in a Phase 2 trial. VIR and Alnylam rapidly recognized the potential of siRNAs to inhibit coronaviruses and quickly expanded our collaboration to include CoV-2 as a target. Working rapidly, Alnylam generated a large number of siRNAs. Working together, the companies have tested the siRNAs in a series of assays and have identified a lead candidate that reduces the level of CoV-2 replication by several orders of magnitude. This siRNA is being moved rapidly toward clinical development for direct administration to the respiratory tract. Our goal is to begin file an IND by the end of the year.

Our third approach is to identify cellular genes whose inhibition prevents infection with CoV-2. The approach uses CRISPR constructs to either inactivate or activate every gene in the human genome in a way that leads to any single cell having a single gene alteration. The collection of cells, which among them have every gene altered, can be infected with SARS-CoV-2. Those cells that are refractory to viral replication or that lead to reduced viral replication can be characterized, and the cellular genes responsible for the virus resistance can be identified rapidly. Using sophisticated machine learning algorithms, the genes can be placed into pathways, networks, and cellular processes. We have done this repeatedly with other pathogens including flu, RSV, rhinovirus, and HBV, and have identified the majority of genes that interfere with viral replication when altered. If we are lucky, we will identify some genes that already have been targeted with compounds. That has happened with some of our other screens, and when those compounds were tested, they have potent antiviral activity. We will know within the next weeks or months whether we are able to identify anti-CoV-2 compounds in time to break this pandemic.

Three significant programs using industry-leading approaches is a lot for a company the size of VIR to take on. Fortunately, we have the help of capable and committed partners with each approach. It has been truly heartwarming to see the amazing response to COVID-19 throughout the industry. In a very short time we have established new relationships with GSK, Samsung, Wuxi Biologics, Generation Bio, and others, and have amended our preexisting relationship with Alnylam. We are close to finalizing a relationship with Biogen as well. We are able to do this because of the commitment of each of those companies to put the rapid development of therapies ahead of normal business concerns. People across the industry obviously recognize the seriousness of the situation and have mobilized to do all that they can, and to do so as quickly as possible. We have a major collaboration with GSK that covers our antibodies, our CRISPR approach, and eventually a vaccine. We have begun working together based on a binding preliminary agreement because both companies recognize the urgency and will complete the definitive contract while the work is going on. Similarly, our manufacturing agreement with Samsung is based on a binding letter of intent, and Biogen has begun working on our antibodies based on a letter of intent while we negotiate the final agreement. We amended our agreement with Alnylam in a matter of days.

These interactions, which in my experience are highly unusual, are a sign of how seriously people around the industry are taking this situation. Our industry obviously suffers from a negative public perception, some of which is self-inflicted and deserved, but much of which is incorrect and based on faulty assumptions. Our response to COVID-19 gives us a chance to demonstrate our true motivations. Many companies, including Vir, began aggressively working on COVID-19 at a time when the economic returns were unclear. To a large extent, the economics of a COVID -19 therapy or prophylactic are still unknown. We can only guess the shape of the epidemic in the future, we don't know to what extent the therapies that arise from our investments will be covered, and of course we don't yet know if they will even work. Despite those risks, our industry has mobilized to do what we can.

We are at a defining moment for our industry. What we are doing, if successful, will not only affect the millions of people who otherwise would have been sickened or died, but will change life for virtually everyone on the planet. This is a once-in-a-lifetime event: years in the future, we will be asking ourselves what we did to bring the epidemic to an end. Our younger colleagues will tell stories to their children and grandchildren. I, for one, do not want any regrets that I left anything on the table—that I could have done more. So many people around the industry have responded in this manner. I have never been prouder of our industry and the people in it. It is time for all who can to step up and act. We have the capability to do something meaningful, and with that capability comes the responsibility to actually do it—and to keep doing it until a solution is found. Working together, we will bring this to an end.

George Scangos has been CEO of three biotechnology companies: VIR since January 2017, Biogen from 2010 through 2016, and Exelixis from 1996 to 2010. From 1986 to 1996, he worked at Bayer, where he had several positions including SVP of R&D and president of Bayer Biotechnology. He serves on the boards of Agilent, UCSF, Cornell University, and the Fondation Santé. He appreciates the amazing colleagues he has known over the years, many of whom contributed to this book.

VIR combines immunologic insights with cutting-edge technologies to treat and prevent serious infectious diseases. VIR has clinical programs focused on hepatitis B and flu and plans to initiate several additional clinical trials for those diseases as well as COVID and HIV in 2020. VIR's reason for being is to combat serious diseases like COVID.

PFIZER AND THE RACE FOR VACCINES AND TREATMENTS

John Young

> "We don't know when the next pandemic might strike but experts do agree that we are long overdue an outbreak on the scale of the 1918 Spanish flu outbreak that killed some 50 million people worldwide."
>
> —Aisha Majid, *The Daily Telegraph*, December 9, 2019[1]

During the lockdown, I read Erik Larson's excellent biography of Churchill's first year as prime minister, *The Splendid and the Vile* (Crown, 2020).

The analogy of the fight against COVID-19 being akin to a war is one being made by many during this unique period. Countless aspects are, of course, very different in the current pandemic from the situation Churchill faced in the opening rounds of the Second World War. Not least, we have the benefit of looking back on World War II with seventy years of perspective and of knowing the outcome. However, I think there are some interesting parallels with the story Larson tells so beautifully.

1. https://www.telegraph.co.uk/global-health/science-and-disease/pandemic-clock-ticking-race-against-time-universal-flu-vaccine

Churchill, in his eulogy for his predecessor, Neville Chamberlain,[2] said: "It is not given to human beings, happily for them, for otherwise life would be intolerable, to foresee or to predict to any large extent the unfolding course of events."

Governments around the world will try to tell their countrymen and women what a great job they have done and how things would have been so much worse without their prompt and decisive actions. As all Monday morning quarterbacks do, their critics, in the inevitable inquiries and investigations that will surely follow, will tell a different story of how unprepared and indecisive our leaders were and why, with the benefit of perfect 20/20 hindsight, our countries should have been more prepared, and should have taken bolder action sooner.

The truth is likely somewhere in between, though I don't intend to address those issues here. I do want to share a personal perspective of what it was like as an executive at Pfizer, the world's leading biopharmaceutical company, to watch this viral battle unfold and how, along with the rest of the world, we began to appreciate its impact and significance for every human on the planet, and to ask ourselves how we might be a part of the solution.

As with a conventional war, we face a dangerous enemy in the SARS-CoV-2 virus with the power to bring death and economic and social destruction on a vast scale.

In this context, *The Splendid and the Vile* is a perfect description of so much that we have seen since the start of 200.

The Vile

As of the time of writing, there were 3,503,533 confirmed cases of COVID-19 and 247,306 recorded deaths globally. In reality, there will be many more undiagnosed cases and unattributed deaths given both the huge variance in testing rates (from over 165,000 tests per million population in the Faeroe Islands, to four tests per million population in Yemen), as well as the variable accuracy of mortality statistics. Both

2. https://winstonchurchill.org/resources/speeches/1940-the-finest-hour/neville-chamberlain

these numbers will be significantly different by the time this is published. We have seen that COVID-19 can affect anyone, with high-profile positive cases including Prince Charles, British Prime Minister Boris Johnson, and actor Tom Hanks.

However, we also see that social determinants of health cause[3] the most vulnerable to be at significantly higher risk, including the elderly and those with underlying health conditions, who are disproportionately likely to be socioeconomically disadvantaged.[4] And those at highest risk include those on the front line helping diagnose and treat patients infected with the virus.

To help "flatten the curve" and ensure that health care systems can cope, governments globally have taken unprecedented measures to implement social distancing, including the virtual shutdown of large sections of the economy. The real economic impact on Main Street as well as Wall Street was immediate and devastating. The Eurozone economy shrank by over 14 percent in the first quarter of 2020,[5] and the US, which was affected a few weeks later, by over 4 percent. Unemployment in the US reached twenty-two million in April,[6] a dramatic increase from a steady state of under six million in 2019.

The economic cost to governments around the world is enormous. In the United States alone, the CARES Act provided $2 trillion of much-needed financial relief, while adding an estimated $1.7 trillion in new government debt over the next decade.[7]

The impact on the social fabric of many countries is becoming evident as societies grapple with the balance of the government's role in

3. https://www.cdc.gov/nchhstp/socialdeterminants/faq.html

4. https://www.washingtonpost.com/opinions/2020/04/26/we-must-address-social-determinants-affecting-black-community-defeat-covid-19/

5. https://www.wsj.com/articles/eurozone-economy-suffers-record-contraction-steeper-than-u-s-11588242162

6. https://www.washingtonpost.com/business/2020/04/16/unemployment-claims-coronavirus/

7. https://www.wsj.com/articles/stimulus-to-add-1-8-trillion-to-u-s-budget-deficit-over-decade-cbo-says-11587072186

ensuring public health and safety for the population at large, balanced by legitimate questions about individual rights and freedoms.

These facts are well known. However, they bear emphasizing because the COVID-19 pandemic illustrates on a dramatic global scale what happens every day of every year; that poor health has a real impact on the real lives of real people.

They also starkly illustrate that, important though hospitals are for patients who really need them, they cannot ever be the most efficient or effective way of managing a disease like COVID-19. For that, we need prevention in the form of vaccines and treatments in the form of antiviral therapies to stop patients with the virus from becoming seriously ill. For patients who do become seriously ill, we need targeted immune-modulators[8] to selectively dampen the immune response that causes severe respiratory damage and appears to be the most common cause of death in patients with COVID-19.

The Splendid

Faced with the greatest health crisis of our time, there is much to be thankful for and to celebrate. The amazing sacrificial dedication of first responders, doctors, nurses, and other health care professionals is something that will live long in our memories. Many others have played a vital role in enabling social cohesion and the simple ability to put food on the table, including delivery drivers, store workers, and many others.

It has been my privilege to work in the biopharmaceutical industry for Pfizer, my employer for over thirty years. In that time, if I have learned anything, it is that great achievements need a great team, as none of us accomplish anything of significance alone. During this crisis, in my role as chief business officer, I lead Pfizer's cross-functional team supporting external collaborations to help combat COVID-19. I have been privileged to be part of an extraordinary team, led by Pfizer's chair and CEO, Albert Bourla, who recognized early in this pandemic that this was not "business as usual."

8. https://www.wsj.com/articles/haywire-immune-response-eyed-in-coronavirus-deaths-treatment-11586430001

In early March, before formal state and federal distancing measures were implemented, Albert laid out three simple, but clear priorities for Pfizer:

1. Protect the safety and well-being of our colleagues.
2. Maintain the continued supply of our medicines and vaccines to patients around the world.
3. Collaborate with experts within and outside of Pfizer to contribute medical solutions to this pandemic.

As the COVID-19 outbreak hit China with increasing ferocity in January 2020, Pfizer established a crisis management team, led by the president of our emerging markets team, Susan Silbermann, to provide support and guidance for our colleagues in China, as well as to coordinate the company's global response. As the pandemic spread, that cross-functional team worked tirelessly to ensure that we delivered on our first priority: protecting the health and safety of our colleagues. As a result of their recommendations, Pfizer promptly issued work from home guidance ahead of government directives. The timely guidance helped ensure that we were able to make decisions on remote working for colleagues who could do so and to ensure that colleagues had the IT support to enable them to be productive.

Our second priority, to maintain the continued supply of our medicines and vaccines, required all our manufacturing plants around the world to stay operational. To do so, 20,000 colleagues in our Pfizer Global Supply organization globally need to work safely at our manufacturing sites. All nonessential colleagues at these sites were asked to work from home, and strict protocols for distancing were swiftly introduced, along with tools for contact tracing.

As growing numbers of patients with serious breathing difficulties began to flood intensive care units, we saw significant demand spikes of over 500 percent for some critical injectable medicines. When patients need to be put on a ventilator, injectable muscle relaxants and pain relief is a vital part of medical protocols so they can be ventilated without distress. As the largest supplier of these medicines to the US hospital system, we knew that the ability of hospitals to help these patients depended on us. Our manufacturing plants at Kalamazoo, Michigan; Rocky Mount, North Carolina; McPherson, Kansas; and Vizag, India, where over 12,000 manufacturing colleagues work, successfully maintained production of these important medicines. In one year, over one billion doses of these vital medicines will go from our manufacturing plants to hospitals in the US and the rest of the world.

I am incredibly proud of these unsung heroes in our manufacturing organization, led so ably by Mike McDermott, president of Pfizer Global Supply, who have demonstrated incredible professionalism and commitment to ensure that we were able to continue supplying these critical medicines at a time when the world has truly needed them.

Our third priority was to be part of discovering medical solutions to the pandemic, and on March 13, Albert announced Pfizer's five-point plan[9] to help scientists and companies across the biotechnology ecosystem bring forward therapies for COVID-19 and prepare the industry to respond more effectively to future health crises:

1. Sharing tools and insights
2. Marshaling our people
3. Applying our drug development expertise
4. Offering our manufacturing capabilities
5. Improving future rapid response

These promises have provided a consistent framework for all our subsequent work and elicited a flood of inquiries from academic institutions and small and medium biotech companies, as well as some industry peers. Some 363 unique inquires asking if Pfizer could help were received in the next forty-five days, each receiving a detailed response and follow-up discussion from a dedicated team, usually within forty-eight to seventy-two hours.

---

9. https://www.pfizer.com/news/press-release/press-release-detail/pfizer-outlines-five-point-plan-battle-covid-19

At Pfizer I am blessed to work alongside some of the world's best scientists and researchers. As the devastating global impact of COVID-19 became apparent, in mid-February, I talked to my friend and colleague Dr. Mikael Dolsten, Pfizer's chief scientific officer, about whether an mRNA vaccine could be a solution to COVID-19.

Pfizer and the German biotech company BioNTech had initiated a partnership in mid-2018 to develop an mRNA vaccine for seasonal flu. This technology injects small sections of mRNA contained in tiny lipid "nanoparticles." mRNA is the genetic material that carries instructions for a cell to make specific proteins, and if you can instruct a cell to make proteins found on the surface of a virus (an antigen), the body's own immune system can mount a response to them and, potentially, provide immunity from any virus with that protein on its surface.

Think of it as a "cell factory." The viral protein produced by the cell can be recognized by the body's own immune system as foreign, and our immune cells then mount a defense response including production of antibodies. The now-trained immune system is ready with a memory function when the real virus enters and can quickly counter it by mobilizing these specific antibodies and immune cells.

This is not just a "cool" technology, but it has the advantage of being able to swap out the mRNA instructions the vaccine contains very quickly and so lends itself to the development of a potential vaccine for new and emerging threats—such as COVID-19.

By the time we first spoke, Mikael was already in discussions with the chief scientific officer of our vaccine research unit, Dr. Kathrin Jansen. Kathrin is an outstanding and experienced vaccine researcher, and she and her team had been working with our partners at BioNTech over the past two years on the basic science required to understand how to effectively design an mRNA flu vaccine. In turn, Kathrin had already held informal discussions with the BioNTech team, who had initiated a COVID-19 research program about three weeks earlier.

The potential for a collaboration to harness the experience and capability of both organizations to rapidly progress a potential vaccine candidate was clear. BioNTech has been conducting research on mRNA for more than a decade, and Pfizer is a world leader in vaccines R&D, with global capabilities in manufacturing and distribution. Normally a partnership agreement will take many months to negotiate; however both parties recognized that speed was paramount. Our business development team worked along with Kathrin, her R&D team, and other colleagues to sign a letter of intent to partner on March 17 to allow work to progress immediately, with the final agreement being formally signed on April 9, only eighty-nine days after the WHO first reported the genetic sequence[10] of the novel coronavirus causing the pneumonia cases in China on January 12, 2020.

This program, BNT162, codenamed Project Lightspeed, received approval from German regulators to begin clinical trials there on April 22, and in the US the first patients were dosed on May 4 after approval from the US FDA. These studies are ongoing as of this writing, and we await data to help ensure that this potential vaccine candidate is both safe and effective. At this early stage of development, the typical probability that a novel vaccine will successfully complete clinical trials and demonstrate it is safe and effective enough to be approved by the FDA and be potentially used in hundreds of thousands or millions of people is only around 10 to 20 percent.

In addition to this vaccine program, back in 2003, Pfizer had developed a number of antiviral drug candidates to help fight the SARS viral pandemic[11] which affected more than two dozen countries and killed about 800 people. Mikael and our research team quickly identified the potential for these protease inhibitor drugs to be active against the virus that causes COVID-19—SARS-CoV-2—and arranged for one of these compounds to be tested in vitro against the virus. As Mikael and his team led by Dr Annaliesa Anderson suspected, the lead drug candidate showed impressive "in-vitro" effectiveness against the virus, and work is currently underway to manufacture enough drug substance for clinical trials to potentially begin in mid-2020.

---

10. https://www.who.int/news-room/detail/27-04-2020-who-timeline---covid-19

11. https://www.cdc.gov/sars/about/fs-sars.html

I don't know if we will succeed in these endeavors. Pfizer has not sought or received any government funding and we are conducting this research, along with our partner BioNTech, entirely at our own financial risk. However, across the whole biotechnology sector, many companies are working alone or in partnership with governments and academic research centers to progress other vaccine and treatment programs. These individually come with significant risk of failure and at a significant cost. But collectively, the probability is high that the biopharmaceutical industry will be able to develop a safe and effective vaccine, effective antiviral treatments, and targeted immune-modulators that patients and the world at large so desperately need.

While we hope that vaccines and treatments will help manage and ultimately defeat COVID-19, we need to be prepared for the virus to potentially be prevalent for decades, and we need the capability to respond to potential new strains of the virus. We should learn from this unprecedented global crisis and ensure that the world has vaccine platforms capable of rapid development and deployment to prevent the human and economic tragedy of COVID-19 ever happening again.

I am a person of faith and as a committed Christian I have never found a contradiction between my faith and the power of science. Many others, including Dr. Francis Collins, director of the NIH, also combine personal faith[12] alongside their work as scientists. Although I am no Churchill, in the same way that in the dark days of 1940 he had a clear conviction that ultimately the war would be won, even though many trials lay ahead, I have great confidence that we can prevail in the ultimate outcome of our battle against COVID-19—and that Science Will Win.

John Young is Pfizer's chief business officer and a member of its executive leadership team. A scientist by training, he has more than thirty years of experience with Pfizer and has held a number of senior lead-

12. https://www.theatlantic.com/ideas/archive/2020/03/interview-francis-collins-nih/608221

ership positions across the organization, including its innovative and essential commercial businesses.

Pfizer applies science and global resources to discover and deliver breakthrough therapies that change patients' lives. We strive to set the standard for quality, safety, and value in the discovery, development, and manufacture of innovative medicines and vaccines. For more than 170 years, we've worked to make a difference for all who rely on us.

CLIMBING A NEW MOUNTAIN: REFLECTIONS ON CREATING A POTENTIAL VACCINE AGAINST COVID-19

Stéphane Bancel

Exposure . . . the thing we are all trying to avoid today. It was actually exposure that ultimately guided me to the position I am in now, leading Moderna through this pandemic. My career brought me in the path of a number of pandemic incidents: in 1996, when I directed sales and marketing at bioMérieux Asia-Pacific during the largest E. coli outbreak ever reported in Japan; and in 2009, as CEO of bioMérieux, when the H1N1 pandemic had devastating economic implications in Mexico.

While I was not sickened by these terrible diseases, I saw firsthand how they can severely disrupt the health and welfare of citizens, communities, governments, and businesses on a global scale. Although today's novel coronavirus outbreak demands new skills, these experiences offered valuable lessons for the current crisis.

Even before I joined Moderna in 2011 as the company's founding CEO, I came to believe messenger RNA (mRNA) could be a technology to address pandemics at scale. Fortunately, there were others at the US National Institutes of Health (NIH) who felt the same way. In 2017, Moderna began collaborating on an investigational MERS vaccine with the National Institute of Allergy and Infectious Diseases (NIAID), part of NIH. As recently as September of last year, we were planning for a pilot pandemic study for early 2020.

Then, everything changed in January 2020. The following timeline recounts key milestones—and some personal reflections—during the sixty-three short days it took Moderna to go from sequence selection to first human dosing of our potential vaccine against COVID-19.

January 2

I was sitting in France enjoying a cup of hot tea while reading *The Wall Street Journal.* I remember watching the sun rise and starkly realizing that this disease in China was not a new strain of flu. It was a coronavirus and we had to chase it! We had a responsibility to use our mRNA platform to help people. It was a moment of realization and commitment that has kept me resolutely focused ever since.

January 13

Just two days after the Chinese authorities shared the genetic sequence of the novel coronavirus, the NIH and Moderna's infectious disease research team finalized the sequence for mRNA-1273, our potential vaccine against COVID-19. Together, we realized important similarities to the MERS virus and, based on good data and analysis from the previous two years, decided to encode for the Spike (S) protein. At that time, NIAID also disclosed its intent to run a Phase 1 study using our mRNA-1273 vaccine in response to the coronavirus threat. We quickly mobilized toward clinical manufacture.

Week of January 20

This week I realized everything would change. I remember seeing the early mortality rates and estimated R0s—the average number of people each infected in turn. I checked the flights going out of Wuhan and thought this outbreak had a high probability of becoming a global pandemic.

January 23

After just two days of negotiations, Moderna announced a new collaboration with the Coalition for Epidemic Preparedness Innovations (CEPI) to develop our mRNA vaccine against the novel coronavirus. We realized that advances in global public health require the collective effort of public-private partnerships—no organization can act alone. It was the financial support from CEPI and the multiyear scientific collaboration with the NIH, along with a multitude of other past collaborations, which helped us adapt quickly to this new scenario.

We also needed to shift minds and priorities internally. The people of Moderna rallied around this new mountain to climb. Our challenge was to show the world how fast we could move, while remaining ever mindful of safety and quality considerations. Our team was energized to demonstrate this.

January 26 and 27

I changed my plans to travel to Holland for a board meeting and instead flew to Washington, DC. There, I met with our contacts at the NIAID, the Biomedical Advanced Research and Development Authority, the Defense Advanced Research Projects Agency (DARPA), and the Food and Drug Administration.

Forging strategic alliances with these organizations early on positioned Moderna as a trusted partner. I reflected on Moderna's early days, when we received a small grant from DARPA in 2013, and how grateful I was for the relatively modest sum, which ultimately helped open doors to the even-more-critical discussions now underway.

February 24

In just forty-two days from sequence identification, we released our first batch of mRNA-1273 for human use. Vials of mRNA-1273 were shipped to NIAID to be used in the planned Phase 1 study in the US.

This was a historic milestone. I was so proud of the Moderna team for their extraordinary effort in responding to this global health emergency in record speed.

March 2

I attended a biopharmaceutical executives meeting with the White House coronavirus task force. This engagement reminded me that, as an industry, our mission calls us to be collectively more collaborative and creative than ever. We had an opportunity to fight this pandemic by building on our successes. We also had an obligation to help ensure that it never happens again.

March 2 evening

The FDA allowed the Phase 1 study of mRNA-1273 to proceed to clinical trial.

March 11

The World Health Organization declared the novel coronavirus a pandemic. In that moment, I realized that science was the only way that we were could solve a problem of this global scale.

March 16

The NIH announced that the first participant in its Phase 1 study of our mRNA-1273 was dosed in Seattle—only sixty-three days from sequence selection to first human dosing. Moderna was the first company to launch a SARS-CoV-2 vaccine trial in humans. It was an extremely exhilarating time and we designed the trial to provide important information about safety and immunogenicity. We actively began preparing for a potential Phase 2 study under our own Investigational New Drug application.

In the days that followed our sixty-three-day journey, the team continued to move at what we now affectionately call "Moderna speed." Exactly one month later, on April 16, we announced an award from BARDA for up to $483 million to accelerate development of mRNA-1273. Time was of the essence to provide a vaccine against this pandemic virus. We had already begun to prepare supply for a second phase trial at our own expense. By investing in our manufacturing process to enable large-scale production for pandemic response, I believed that we could supply millions of doses per month in 2020 and, with further investments, tens of millions per month in 2021. This all assumed, of course, that the vaccine candidate proved successful in the clinic.

To enable larger scale manufacture of mRNA-1273, we announced a ten-year strategic collaboration agreement on May 1. I was pleased to partner with Lonza, which shared our commitment to rapidly addressing the pandemic. The agreement will enable Moderna to accelerate, by ten times, our manufacturing capacity for mRNA-1273. Lonza's global presence and expertise will be critical as we scale at unprecedented speed and potentially enable manufacturing of up to one billion doses per year of our vaccine against COVID-19.

At the time of this writing, we continue to scale up our manufacturing capabilities while our vaccine is tested in the clinic. In parallel, our paramount obligation is to ensure the safety of all participants in our ongoing clinical programs and the integrity of the studies in which they participate, as well as the safety and well-being of our employees.

We plan to start a pivotal Phase 3 in early summer 2020.

Our remarkable team of more than 900 colleagues is the engine behind everything we have been able to accomplish at Moderna. As I reflect on the experience overall, I believe it has motivated us personally to perform at levels we may never before have imagined. I am forever thankful to my colleagues for their commitment to advancing our technology and to creating and manufacturing a potential mRNA vaccine against COVID-19.

Moderna people are extremely proud of the work we are doing. Unlike with some diseases, almost everybody has been touched by this virus. Yet, survival is in our DNA. We fight to get things done in extreme moments. Most of us would not know we had this in us before it all happened.

Our journey is far from over. In my mind, it is our responsibility as an industry, as a company, and as global citizens to help ensure a pandemic of this magnitude never happens again. I've worked in infectious diseases for my entire career—and I've been exposed to the disruption, the health crises, and the economic fallout they create. It is up to all of us to bring forward solutions. I hope you will join the many companies, worldwide health agencies, and nongovernmental organizations that are answering the call. Together, we can all use our skills to help serve people in this remarkably urgent time of need.

Stéphane Bancel is chief executive officer and a member of the board of Moderna Inc. He is a director at Qiagen NV, a venture partner at Flagship Pioneering, and a trustee of the Museum of Science in Boston. He also serves on the board of advisers of Life Science Cares.

Moderna Inc., a clinical-stage biotechnology company, develops therapeutics and vaccines based on messenger RNA (mRNA) for the treatment of infectious diseases, immuno-oncology, rare diseases, cardiovascular diseases, and autoimmune and inflammatory diseases. These mRNA medicines are designed to direct the body's cells to produce proteins that can have a therapeutic or preventive benefit. The company's platform builds on advances in mRNA science, delivery technology, and manufacturing, providing Moderna the capability to pursue a robust pipeline of new development candidates, independently and with strategic collaborators. Moderna has twenty-three development candidates and has started clinical trials for sixteen medicines. The most advanced is planned to enter Phase 3 in early summer 2020.

FIGHTING THE WAR AGAINST RNA VIRUSES: REPORT FROM THE TRENCHES

Jean-Pierre Sommadossi

I am a scientist and an entrepreneur. Most of my career has been dedicated to discovering and developing drugs to combat serious viral diseases. I clearly recall the first days of HIV, when we had no idea what that virus was, where it came from, how it was transmitted, or what we needed to do to fight it. All we knew is that it was killing lots of people and we needed to find a way to fight it. As a KOL at that time, I remember the first direct-acting antiviral (DAA) drugs, like AZT and DDI. I remember the early days of the combination therapy debate. I remember that, when we first proposed a clinical protocol of a triple-combination of nucleoside analogs as a cocktail, such a study was possible only as an international investigator-initiated trial, as no pharma company was inclined to take the risk. It did work and led the way to the cocktail that now enables people with HIV to live long and healthy lives. HIV is no longer the death sentence that it was in 1985; we won. It was less than forty years ago, but it seems like a millennium. The key players in antivirals at that time were a small club: Burroughs-Wellcome, BioChem Pharma, Agouron, and later on, Gilead, Vertex, and Triangle Pharmaceuticals. The biotech companies were the cornerstones of the fight against HIV.

After that, things became both easier and harder. Hepatitis B and hepatitis C are devastating chronic diseases, and the development of treatments for each has taken another decade. And we won again. Gilead, as well as Idenix and Pharmasset (which I cofounded) led the war against hepatitis C. Sovaldi is currently the miracle drug that treats this viral infection that can lead to fatal liver damage and cirrhosis.

Most viruses that have caused havoc in our lives are single-stranded RNA viruses. The list is long and includes such household names as HIV, hepatitis C, Zika, dengue, chikungunya, viral encephalitis, Ebola, and, not surprisingly, the multitude of coronoviridae culprits, led by SARS, MERS, and now SARS-CoV-2. They are actually very simple creatures, and their only desire is to procreate using our genetic machinery. Once they get inside the host cell, they need to multiply their RNA to make more of themselves. For that, they need a special enzyme, called RNA polymerase, the key that opens the magic box. RNA polymerases are highly conserved, with homology as high as 99 percent in some cases. While the virus can mutate and change many parts of its outer envelope, thus escaping the immune system and making vaccine development difficult, RNA polymerase is its Achilles' heel. Most of the available effective drugs to combat RNA viruses are inhibitors of RNA polymerases, including sofosbuvir (Sovaldi), tenofovir (Viread), abacavir, 3TC, and more recently, remdesivir.

For the past five years, my goal has been to discover and develop the next generation of DAAs that can efficiently and safely combat RNA viruses. Atea was born when sofosbuvir was becoming a reality, but most of us in the field knew that we needed to stay vigilant and deliver even more potent DAAs. AT-527 is the perfect example of the uniqueness of Atea's platform in discovering highly innovative nucleotide therapeutics. This oral RNA polymerase inhibitor is a complex double prodrug synthesized as a specific salt form, which required new breakthroughs in our knowledge of medicinal chemistry and cellular pharmacology. Clinical studies in 2018 and 2019 confirmed the safety and potent antiviral activity of AT-527 in HCV patients. We plan to enter Phase 2b HCV clinical studies in the near future.

I recall being in France to visit my family at the end of February. My family lives on the border of France and Italy, in a beautiful town called Roquebrune-Cap-Martin. The news from Italy were disturbing. I realized at that point that we may have a problem on our hands, though I certainly did not anticipate its magnitude. What was going through my mind was this: more than 20 years ago, our team at Idenix synthesized a number of RNA polymerase inhibitors that we tested against a few human coronaviruses. The compounds were active and we filed a patent; these compounds were first-generation DAAs and not as efficient as what we have today, but it was a start.

Remembering these findings, coupled with a decade of breakthrough research into the mechanism of coronavirus infection, I asked one of our scientific collaborators to test AT-527 against a series of human coronaviruses. As I had hoped, the drug was quite active. The Atea team then contacted the NIAID, which validated the data. We then knew that we had a potent oral DAA that could be used on the front lines of treatment for COVID-19. The decision was made to shift focus from our original pipeline and to concentrate our near-term resources on COVID-19.

Just a few words about our extraordinary team. We are ten individuals who have worked together for over 20 years. Our expertise spans all critical phases of discovery, preclinical, and clinical development of small molecules, and we have been able to master (several times) all these phases in record time within the budget of a small biotech company. In less than a month, we assembled a full IND package, which was submitted directly to the FDA, without a pre-IND meeting.

I am not sure how many people reading this understand the process of FDA review necessary to initiate a clinical trial. We received the "ready to proceed" notification letter from the FDA in less than twenty days, after three detailed FDA reviews and responses with two-day turnarounds, implying fourteen- to fifteen-hour work days, seven days a week for our team. Such outcome within such short time is unprecedented in my experience, and I cannot thank the FDA more for the speed and diligence with which they responded to our application

So what now? We started our Phase 2 clinical trial in late May in moderate COVID-19 patients. I believe that our drug will ultimately be used as a frontline antiviral agent that will slow viral replication, which will allow the person's immune system to recover enough to get the infection under control. Since AT-527 is oral, we plan to expand our clinical studies to the health care workers who are on the front line of the battle against this terrible disease and then move quickly to an outpatient community setting.

I was thinking of how much has changed from the early HIV days. This virus is nasty, tricky, and lethal. How much have we learned in such a short time—just a few months? Not as much as we need, but enough to get a good start. I am certain that if this virus had presented itself forty years ago, our future would have been very bleak. Today, we have better tools, multiple drugs, and vaccines in development, as well as an industry that has grown to be well funded and well equipped to tackle some of the most lethal human diseases. We will win again.

Jean-Pierre Sommadossi has spent his career battling viruses. He has built and lead companies that have successfully tackled hepatitis C, B, and HIV/AIDS, creating landmark therapeutics along the way. He is committed to world-class science, has authored over 150 papers, and holds over sixty patents.

Atea is a biopharmaceutical company discovering and developing best-in-class therapies to address life-threatening viral diseases. In March 2020, the company pivoted all its efforts to develop an oral direct acting antiviral therapy for COVID-19.

BEYOND THE CUL-DE-SAC

John V. Oyler

As a young boy, I would run around with neighborhood kids in the woods near my home in Pittsburgh. I was excited to make my first best friend and utterly heartbroken when he moved away because his father lost his job in the recession. Stress gnawed at his parents and deeply affected my friend. I had similar experiences with my second and third best friends. I vividly remember lying in bed dreaming of a place where friends could work together at good jobs and live as neighbors on a cul-de-sac. I understood how one person having a wonderful job, a bad job, or no job, affected an entire family and those around them. I believed there could be nothing more meaningful than creating jobs to help families and their communities.

After an otherwise blessed childhood, I experienced much greater loss as an adult. Three of my closest, most vibrant friends passed away. In a moment of clarity, I made the fight for life the singular focus of my career.

A decade ago, I met my BeiGene cofounder, Xiaodong Wang, who is an exceptional biochemist, one of the youngest ever US Academy of Science members, and a wonderful man. We shared a vision of forming a team of exceptional, passionate people, wherever they might be, to work with a clear purpose. We decided to leverage science, logic, and worldwide collaboration to create novel and affordable medicines. To be affordable, we believed that we must address the single biggest issue for our industry: the time and money spent on lengthy human clinical trials. We believed that regulatory changes enabling a single global trial to replace multiregion trials soon would occur, profoundly reducing cost and time. Part of our vision was to build a unique internal capability to run high-quality global clinical trials geared toward this new era. We believed that BeiGene could have a transformational impact by accelerating the development of innovative medicines and making them available and affordable to billions of patients worldwide. Our first effort would be fighting cancer.

BeiGene is now a global oncology company comprising thousands of people on four continents operating in over thirty countries, discovering, developing, and marketing medicines. Our greatest staff concentrations are in the world's two largest health care markets: the United States and China. Our mantra is, "Cancer has no borders. Neither do we."

From inception, BeiGene has been a borderless operation. Well before the world sheltered in place, we frequently said that our headquarters were on Zoom. We believe that centralized headquarters sends a message to all outside that they are second class. We strive to include ideas and talent from everywhere and to avoid internal politics that might interfere with allowing science and logic to lead the way. When we find the right people to execute our vision, we let them flourish where they are rooted. Anchored but borderless thinking has helped us build a cancer medicine portfolio through internal discovery and partnerships. It has also given us the tools to work remotely and the nimbleness to adapt quickly to the world evolving around us.

BeiGene has been united against cancer, a devastating enemy that attacks people of all ages, from all countries. When we learned of the looming threat of COVID-19, we understood immediately its potential to cause catastrophic global damage. With an office in Wuhan, we witnessed early on the havoc this virus was causing. I knew we had to join the fight against it. Since we are not an antiviral company, it was clear that our role was not to lead the charge but to be a team member, connector, information provider, and contributor.

Key concepts that have been meaningful in our evolution to date, and now through the COVID-19 crisis, include:

Passion and sense of purpose

One of our team members recently observed that as we grapple with the challenges of COVID-19, we feel vulnerable and uncertain, and are reminded of our own mortality. This, he said, is what cancer patients and their families face every day. This is why we fight so hard for life. I am proud of the ways in which BeiGene's teams have come together over the last decade to fight cancer, and now to assist in the fight against COVID-19.

Preparedness to deal with uncertainty

By virtue of starting from nothing and trying to do things differently in novel medicine discovery, we are accustomed to facing risk and uncertainty every day. By early January, we anticipated threats to our team members, so we moved quickly to start closing offices and eliminating travel. We shifted even more to using Zoom to get work done, and remained focused on fulfilling our commitment to cancer patients while keeping our people and their families safe. With new workstreams related to COVID-19 and quarantines, we needed to adjust dramatically, helping and covering for each another. We have urged associates to ask for help when they need it. I tried to be there for the team, and at each turn I was surrounded and supported by good friends with a shared vision.

Understanding that no challenge is too big

Although it would have been easier to just write a check to support frontline responders, we worried that cash could take too long to have an impact and might be misdirected. Instead, we navigated daunting logistics and, by January 24, shipped our own protective gear directly from our manufacturing facilities to some of the most inundated Chinese hospitals. Simultaneously, we were sourcing needed gear for frontline responders in China, and subsequently for responders on three additional continents. Team members and their families contributed both financially and logistically to these efforts.

Leveraging our strengths

We called on strong relationships in China and the United States to facilitate communication among industry contacts. This helped US hospitals sourcing supplies from China, biotech executives connecting with each other, and third parties sourcing and transporting ventilators to New York City hospitals.

Anticipating future possibilities, good and bad

In February I alerted my colleagues on the Biotechnology Innovation Organization's board of directors to the seriousness of the looming pandemic threat. I urged them to act quickly on their plans to coalesce a response from the biotech industry, and they have risen to the challenge. We also understood the burden this placed on our team.

Communicating openly and frequently

Our leadership team hosts frequent Zoom town halls for the company, regions, and groups. We provide multilingual channels and simultaneous translation, and we answer all questions raised. Once our teams had a chance to adjust to the modified rhythm and routine of teamwork in quarantine, ideas started flowing about how we might adapt our use of technology to advance needed collaboration with others in our industry.

We don't give up

We are committed to continuing to fight cancer. We are hiring, investing, and collaborating. I am proud of the way BeiGene's teams have come together over the past decade to fight cancer and now to assist in the fight against COVID-19. And I would be remiss if I did not add that I am deeply grateful to and awestruck by my colleagues throughout the biotechnology industry as well as the scientists, clinicians, first responders, and essential workers who are helping us through this difficult time.

The devastating effects of a contagion like COVID-19 are global by nature. Our responses as communities and as an industry must be global to succeed. Our responses must be guided by science and logic, and not divided by politics or regionalism. Together, we can respond quickly, efficiently, and as affordably as possible. In that spirit, we have found allies beyond our internal teams.

- We joined a three-party collaboration to pursue a novel modality of treatment for COVID-19. One collaborator is contributing its proprietary discovery platform, another is prepared to use its technology platform to develop and manufacture this unique type of antibody, and BeiGene will provide global clinical development support. Given the urgency, we began working together based on a virtual handshake. The entire focus is on evaluating this option as quickly as possible.
- We also have responded to a request from an eminent doctor who has been helping us evaluate our internally discovered medicine Brukinsa (zanubrutinib) for its potential to treat an additional type of cancer beyond its initial indication. This doctor has noted that anecdotal data and a reasonable mechanistic hypothesis suggest that compounds from this target class might also help patients with COVID-19-related pulmonary distress. We are testing this hypothesis. Medicine development is fraught with myriad risks and opportunities to fail. We don't know if this will succeed, but we agree it is a hypothesis worth exploring.

The optimism I felt when I set out to create a different sort of company—transformational, global without a headquarters, lasting, impactful, and involving spectacular science that could help people around the world—is similar to what I feel now. Industry-wide global collaboration holds the promise to advance medical innovation at a pace previously unimagined. Diseases and infections disregard arbitrary boundaries. I've realized that we don't need a meeting, a headquarters, an office, or even a cul-de-sac to be close. United, we can accelerate the development of innovative, affordable medicines to the benefit of billions of people worldwide.

John V. Oyler is BeiGene's chair, cofounder, and CEO. He previously was chief executive at three health care companies and founded Telephia Inc. He began his career at McKinsey & Company. Oyler serves on the board of directors of BIO. He received his BS from MIT and MBA from Stanford University.

BeiGene Ltd. (Nasdaq: BGNE; HKEX: 06160) is a global biotechnology company. Its 3,800+ employees are developing novel cancer medicines to improve outcomes and access for patients worldwide. It markets internally discovered oncology products in the US and China, and also markets or plans to market in China in-licensed oncology products.

UNPOPULAR DECISIONS AND LAME-ASS FOLLOWERS

Paul Hastings

"Boss, you don't want to be accused of being a lame-ass follower." Anybody who is even vaguely acquainted with me would know that such an exhortation might, well, get under my skin. Follower I am NOT—ever—and particularly not while doing my best to chart the treacherous waters of COVID-19. But one afternoon in mid-March 2020, our chief scientific officer, and my good friend, entered my office, looked me anxiously in the eye, and said just that. After I counted to ten, caught my breath, and gasped in amazement, I leaned in and said, "Well, what would you have me do differently?" I had just congratulated our scientific founder for flying to our board meeting from Singapore, and told him I had just come back from a board meeting in Tampa, but all our board members but he were calling in to this meeting. Our CSO took issue with this bravado, and whacked me upside the head . . . yep, well deserved and, boy, did it catapult me into action.

When I accepted to serve as CEO of Nkarta, a cell therapy company developing engineered and off-the-shelf cancer therapies using natural killer (NK) cells, the job description seemed to have left off the part about managing through a world turned upside down. Pandemic is not in the CEO playbook, nor should it be. But it is a leadership challenge that we as an industry must address. It is our moment. It is our moment to rise to the occasion with the science and discoveries that will restore our way of life. It is also our moment to lead by example.

As members of the biopharma industry, we tend to be particularly comfortable with the inevitability of risk-taking. We are accustomed to weighing scientific facts in the face of uncertainty and incomplete information. We regularly adjudicate complex risk-benefit algorithms as we fulfill our responsibilities to patients, employees, and stakeholders. We do not shudder when faced with the potential of failure, but see it as an opening for success, innovation, and value creation. We may not always get it right, but risk adjustment is what we do all day long. And when we do get it right, we've made a difference in the world. In this respect, I think we as an industry are uniquely equipped to be models of fair action and good judgment to our communities. Over time, our collective decision-making practices have led to important rewards for our society. Let us not stop now.

I don't pretend to have the formula for what it means to lead in a historic world upheaval. What follows are some striking experiences of the past few months and examples of leadership during trying times.

- It's March 3. Most of us are still processing the startling news about the first reported cases of community transmission in Seattle. I'm attending the Cowen Healthcare Conference in Boston along with our leadership team. The day before, Peter Kolchinsky and RA Capital kindly loaded us up with bottles of hand sanitizer. We embraced the elbow bump throughout the event, wiped down surfaces in our meeting room, and washed our hands regularly. We enjoyed lively and thoughtful one-on-one discussions with investors about the progress of our pipeline and our excitement about the potential of natural killer cells. Time well spent. However perhaps the most memorable meeting was with an investor who we know and like. In fine spirits, he was struggling with a dry cough during our thirty-minute meeting. As we wrapped up, he noted that a student at his child's school had just tested positive. Just a few days later, the conference organizers would contact us regarding attendees at the conference who had since tested positive. My husband, Steve, reminds me that I'm an at-risk individual by age and chronic disease, and I am lucky, given all the cases from Boston that week, that I made it back to San Francisco symptom-free. Life has clearly become more complicated.

- A few days later, I'm traveling to Tampa to attend a board meeting. I'm in one of those executive passenger buses originally intended to shuttle crowded travelers and their baggage between the airport and car rental, but with a busload of employees of the company I am on the board of. Tight quarters. Sitting next to me was former New Jersey Governor Chris Christie. It was still early days in the COVID-19 saga. We got to talking. He openly reflected on his actions during Hurricane Sandy in 2012. He had made the series of self-described "unpopular" decisions to close casinos, order mandatory evacuations across large swaths of the state, block bridges, tunnels, the Garden State Parkway south, and eventually "postponed" Halloween across the state. For him, unpopular decisions come with the territory. "You even hugged President Obama," I remind him, and tell him how touching that was to me, and how real he was to do that, given the political discourse at the time. Now, he told me, in 2020, a massively unpopular decision on a national scale was urgently called for. He had little doubt about the path of Hurricane COVID-19. In his view, it was time to shut down all nonessential activities across the country. It was time for officials to take unpopular decisions. He saw no other options, and he was going to share that with the commander-in-chief later that day on a phone call.
- March 13. The impact of COVID-19 was rapidly evolving. In the course of twenty-four hours, the quarterly Nkarta board of directors dinner and all-day meeting abruptly pivoted into virtual events. Seasoned directors, road warriors who just days earlier had professed to being undaunted by an infectious disease, informed us one by one that they would attend by videoconference. A single scientific adviser, our founder, chose to travel from Sin-

gapore. That same day, Nkarta put in place our first COVID-19 safety policy. We closed our offices and labs to nonessential visitors and advised all nonessential employees to plan to work from home going forward.

- The Nkarta team has done a remarkable job of staying ahead of the curve. And so has California. San Francisco Mayor London Breed, in a surprise move—one that took amazing courage and thought—ordered businesses closed and issued shelter-in-place policies on March 17. Governor Gavin Newsom followed suit a few days later, with overlapping statewide orders. By the time these orders were handed down, Nkarta was already iterating on its policies to ensure employee safety and support business continuity. Adapting to the new government orders at that point became a nearly seamless process. I continue to be proud and amazed at how our progressive and "sanctuary" state has handled this crisis. Hats off to Mayor Breed and Governor Newsom.
- The strength and consistency of our normal course of business planning, practices, and decision-making put us in good stead as the force of the outbreak hit. We were not starting from square one. For example, we had an outstanding technology infrastructure in place when the shelter-in-place orders came down. We had completed an enterprise-wide rollout of Zoom in January, providing us with a powerful tool to hit the ground running.
- On March 16, we hosted our first ever all-hands meeting via videoconference, where we rolled out an updated COVID-19 safety policy. We continue this weekly practice. In this all-hands setting, I explained some of my own views around dealing with adversity and how my personal history has shaped me into what I called a realistic optimist. I do not minimize the gravity of the tragedy we are living through. However, I believe we must look outward, pursue solutions, and explore possibility, and not wallow in regret and hopelessness.
- March 16 also marked the start of the daily Nkarta leadership meeting. Every afternoon, at the end of our virtual workday, our executive team checks in on Zoom. The cadence of the daily

meeting and its open flexible structure go a long way to maintaining close coordination and strong communication. This practice has been invaluable.

- A key factor behind our successful adjustment has been the strong organizational foundation we had going into the COVID crisis. We've worked extremely hard over the past few years to shape a company culture deeply rooted in transparency, inclusion, authenticity, and high levels of trust. We communicate broadly and regularly with the teams, and share the facts to empower decision-making across the organization. These behaviors have allowed us to be smart and exceptionally nimble as we responded to the unfolding situation.
- Despite the distractions and inefficiencies of remote work, skeleton lab schedules, and social distancing, I continue to be amazed by the commitment and productivity of our teams. Activity levels remain extremely high and we continue to make excellent progress during this challenging period.
- Despite COVID 19 milestones: I hired two awesome independent board members to join Nkarta's board. I actively conducted a search for a new CEO of the Biotechnology Innovation Organization (BIO) with my friends Jeremy Levin and John Maraganore, a full-time endeavor, all done on Zoom! Nkarta is filing an request for an investigational new drug, we are building a clinical GMP manufacturing facility, and we have made a number of key hires. . . . We are not letting a "settle from home" order slow us down, and we are doing it with entrepreneurial spirit and with safety in mind for all our team members.
- We continue our efforts to build upon this foundation and strengthen the Nkarta culture. As part of our process for back-to-work planning, we have actively solicited input through company-wide electronic surveys and have shared these survey results as part of an ongoing two-way dialog.
- A fair, open, and transparent culture also requires space for dissent, debate, and open discourse. I expect our team to bring their ideas to the table, speak their minds, and argue their perspec-

tive. This is the definition of organizational excellence for me. It brings out the best in all of us, myself included.

This last point returns to me to the fate of my friend who warned me not to be a "lame-ass follower." He'll never say that again, because once is enough for me. I aim to please and not to disappoint. He's a critical member of the team whom I continue to admire and adore. That galvanizing exchange only deepened my respect and admiration for him. I believe he understands that I require neither acquiescence nor agreement from my colleagues, only open communication. I do want open, honest, and direct, always! And this was what he was doing. My position was that we must do all that we can to ensure the safety of employees who were carrying out essential activities in the lab. His position was that there were too many unknowns as the dangers of COVID-19 were only beginning to unfold. If we kept the lab open, some employees might feel pressure to work. I asked him to elaborate on what we could do to ensure that everyone felt safe and alleviate any pressure to work. We exchanged ideas. We listened to one another. We found a way forward. We didn't care about popular decisions, just the right ones.

Paul Hastings is the chief executive officer of Nkarta, a privately held biopharmaceutical company developing engineered natural killer (NK) cell therapies to treat cancer. After beginning his career in sales and marketing at Hoffman-LaRoche, he has led more than a dozen biopharma companies as CEO, president, or director. He currently serves as vice chair and member of the executive committee of BIO and is chair and CEO of patient advocacy organization Youth Rally Inc. (www.youthrally.org).

Nkarta's mission is to discover, develop, and deliver novel off-the-shelf natural killer cell therapy product candidates that have a profound effect on patients. Nkarta focuses on combining its NK expansion and cryopreservation platform with proprietary technologies to generate an abundant supply of NK cells, enhance tumor cell recognition, and improve cell persistence and activity for the treatment of cancer.

COVID-19, CANCER AND CAR T: CANCER DOESN'T STOP DURING A PANDEMIC—SO WE CAN'T EITHER

Christi Shaw

The COVID-19 pandemic put nearly every aspect of everyday life on hold. It shuttered storefronts and restaurants, emptied classrooms, and cast an eerie quiet over city streets. People living with cancer are among those most vulnerable to severe COVID-19 symptoms, and the need to protect these patients from contracting the virus—coupled with extreme pressure on existing hospital and health care system resources—has created new uncertainties and fears for patients.

But while a pandemic brings everyday life to a halt, it doesn't slow the progression of cancer, and we cannot wait until the pandemic passes to bring potentially lifesaving treatments to those who need them most.

The rapid spread and stark mortality rates of COVID-19 have made the world acutely aware that we must do more to protect vulnerable populations—not just during a global health crisis, but every day. In the hurdle to bring COVID-19 vaccines and treatments to market quickly, we're also witnessing that it is possible to rethink some processes and protocols, especially when we all come together to work toward a common goal.

The lesson from this experience can be applied to the immediate crisis as well as to our goals as biopharmaceutical companies beyond COVID-19: to move mountains for patients, we need to push past what has been accepted as "the normal" way of doing things.

At Kite, the patients who rely on our CAR T therapy are among the sickest. Most have exhausted all other treatment options, and they do not have time to wait to be treated until the pandemic lifts. Therefore, we must do everything possible to bring our treatments safely to them. In normal times, this means drawing blood from a patient at an authorized treatment center (ATC), separating out the T cells, and shipping those T cells to a manufacturing facility. There, the T cells are engineered into CAR T cells and then are shipped back to the ATC. The patient undergoes three days of low-dose chemotherapy before receiving their infusion at an ATC. After the infusion, patients must be monitored and need to remain near the hospital; if side effects occur, hospitalization may be required.

The knock-on effects of COVID-19 have created many challenges to the process of treating patients with CAR T therapy. It was clear early in the crisis that we needed to make adjustments and mitigation plans and take other measures to protect these individuals—along with their caregivers, health care providers, and our own employees involved in the manufacturing and delivery process—and also ensure that they receive their treatments in time.

Adapting our own carefully crafted processes to create a Plan B—and later, as new challenges emerged, Plan C, Plan D, and so on—has been difficult. Now, in addition to maintaining the highest standards for safety throughout the cell extraction, CAR T manufacturing, and infusion processes, our plans must account for many external and unexpected logistical factors that—if unaddressed—could affect the timely delivery of treatment to patients.

What do you do when hospitals face unforeseen capacity challenges and are unable to accept a CAR T patient?

What do you do when the reliability of air travel, the fastest way to transport a patient's cells, is significantly impacted?

What do you do when the required resource supply is limited?

Of course, patients don't have time for us to wallow in the complexity and enormity of the challenges at hand. Now is the time for fast, creative, and proactive thinking from teams across Kite—and the larger biopharmaceutical and health care ecosystem—to bring actionable solutions to these challenges.

In the weeks since the pandemic took hold, resilient teams at Kite implemented necessary adaptations to our CAR T cell manufacturing and delivery process. New shipping routes and methods to account for disruptions in air travel—in the US, in Europe, and between—were devised. It has been necessary to stay in constant contact with suppliers and vendors to monitor our supply chain and shift resources accordingly. If an ATC is unable to conduct an infusion, we've committed during this situation to not only continue to manufacture the patient's CAR T cells, but to also securely store those cells until the patient can safely visit their ATC. Additionally, we are temporarily helping to inform individual ATCs about alternate ATCs in their region that are in a position to treat patients with CAR T. All this is done to avoid any patient from having to wait a moment longer than necessary to receive their treatment because the Car T patient does not have the benefit of time.

All these decisions—and more—have taken us outside of our "normal" way of operating. They had to be made quickly and adapted as new information about the virus and containment measures came to light. All these decisions were absolutely critical to protecting and enabling treatment of vulnerable patients, and new challenges could call for more decisions tomorrow. Plans change, but our focus on patients never wavers. If something stops working, we will try another way that still allows us to safely get therapy to patients.

This must become our mantra, not just for CAR T, but for the entire medical community as we look to a future beyond COVID-19. We must keep seeking out opportunities to enact positive disruption, and we must try everything we can to make it possible for all eligible patients to access CAR T therapy—whether there is a pandemic or not.

Christi Shaw is the CEO of Kite and also sits on the Gilead Leadership Team. She has full accountability for all aspects of cell therapy at Kite. Her leadership throughout her career spans a broad range of therapeutic areas, including oncology, cardiovascular, respiratory, immunology, infectious disease, neuroscience, ophthalmology, and medical devices.

Kite, a Gilead Company, is a biopharmaceutical company based in Santa Monica, California, engaged in the development of innovative cancer immunotherapies. The company is focused on chimeric antigen receptor and T cell receptor engineered cell therapies. For more information on Kite, please visit www.kitepharma.com.

SCIENCE AND INNOVATION ARE THE ANSWERS

John Maraganore

Ultimately, the answer to this pandemic will be science and innovation. Period. We will need not just one, but an arsenal of therapeutics and vaccines to overcome this global health crisis. While frontline health care workers are bravely working to treat the flood of patients and minimize the loss of life from COVID-19, scientists and clinical trialists are working round the clock to develop and test the new treatments that will ultimately allow us to emerge from this pandemic. When this current phase is over, and we've reached our new normal, it will be transformative science that will have saved scores of lives around the globe.

This unprecedented pandemic has changed the way we live, work, and interact with one another. On this dimension, I am amazed by the ingenuity, speed, and collaboration I've seen across the biopharma industry as we work united by a common purpose to respond to perhaps the greatest public health threat of our lifetimes.

Given how fundamentally COVID-19 has disrupted the structure of our society, it's hard to believe that less than five months ago many of us were at the annual J.P. Morgan Healthcare Conference and started discussing the initial reports about a virus coming out of China. At the time, the "Wuhan Outbreak" seemed a noteworthy, but still distant, threat to global health. At the same time, Alnylam scientists were also following the outbreak, and became interested in exploring the potential of RNA interference, or RNAi, as a treatment to combat the novel coronavirus emerging from Wuhan, which later became known as SARS-CoV-2.

RNAi uniquely suited to address SARS-CoV-2

Over the past eighteen years, Alnylam has led the translation of RNAi into a whole new class of innovative medicines, resulting in two approved therapeutics, two in registration, and many more in late-stage and early-stage clinical testing. Based on Nobel Prize-winning science, RNAi is a powerful, natural cellular mechanism that results in the specific degradation of mRNA as a means to regulate protein expression. As coronaviruses have RNA genomes, RNAi can be used to directly target and degrade the genome of SARS-CoV-2, a unique and powerful antiviral approach.

Interestingly, many of the early RNAi programs were directed at viruses. However, these early programs used naked, unmodified, or only partially modified small interfering RNAs (siRNAs)—the molecules that mediate RNAi—which were susceptible to degradation and induction of immune responses, and as a result, did not lead to effective medicines. Since then, our scientists have made significant advances in the chemistry of siRNAs resulting in stable, immunosilent, and potent compounds.

Further, since siRNAs act inside cells, efficient intracellular delivery to the host cells and tissues is required to elicit an effective antiviral response. Advances in our delivery technology at Alnylam have led to improved distribution and productive delivery to target tissues, initially to the liver, and later to the CNS and the eye. And more recently, our scientists made significant advances in the lung delivery of siRNAs, right around the time we started to hear initial reports about a novel coronavirus causing a mysterious lung disease.

With the course of the coronavirus outbreak so uncertain, and given the potential role that RNAi could play in combating it, we decided in late January to initiate a COVID-19 therapeutic program.

Partnership with Vir Biotechnology

As we began work to design and synthesize a set of over 350 siRNAs targeting highly conserved regions of the SARS-CoV-2 genome, we recognized the value that a partner with complementary capabilities and expertise could bring to the program. Fortunately, we didn't have to look far to find an ideal partner. Vir Biotechnology is an impressive young company focused on infectious diseases, with cutting-edge technology and fantastic leadership, and with which we already have a highly productive existing partnership. Alnylam first entered a collaboration with Vir in 2017 to develop siRNAs to treat infectious diseases, with an exciting program in HBV currently in Phase 2 clinical testing.

Within days of placing the first call to Vir CEO George Scangos to discuss the possibility of collaborating on this effort, we had our research teams talking to each other about the program, and by early March we announced an amendment to our existing agreement to include the development and commercialization of RNAi therapeutics targeting SARS-CoV-2. The collaboration would leverage our expertise in RNAi technology, including the recent advances we had made in lung delivery of siRNAs, coupled with Vir's infectious disease expertise and established capabilities. While we jointly began work on SARS-CoV-2 targeting investigational RNAi therapeutics to treat COVID-19, just over a month later, in early April, we announced yet a further expansion of the collaboration to include up to three host factor targets that are critical for supporting viral infection.

Fortunately, we are seeing examples like this across our industry, with companies coming together to contribute their unique capabilities to accelerate solutions to address the pandemic.

The path forward

Working with our partners at Vir, within just three months after making the decision to initiate our COVID-19 program, we have now declared our first development candidate. The candidate, ALN-COV (VIR-2703), has shown excellent potency, with an effective concentration of 95 percent inhibition (EC95) of less than 1 nanomolar. Further, as the compound was designed to target highly conserved regions of the SARS-CoV-2 genome, it has a predicted reactivity against over 99.9 percent of the viral genomes currently available in public databases. We will now coordinate with regulatory authorities with a plan to initiate human clinical trials at or around year-end 2020.

We plan to advance ALN-COV via direct administration to the respiratory tract (e.g. via intranasal and/or inhalation), the site of viral infection and replication, for both treatment of infected patients and prevention of infection in at-risk uninfected people.

Should ALN-COV prove safe and effective as a treatment for COVID-19, I believe it could potentially have a role not only in this current pandemic, but potentially against a future novel coronavirus outbreak, as the molecule was designed to target highly conserved regions of the SARS-CoV-2 genome and has full reactivity against SARS-CoV, the virus responsible for the SARS outbreak of 2002–2003. More broadly, demonstration of utility to treat COVID-19 will lead to further interest in RNAi technology to counter emerging epidemics, as the technology offers the potential to rapidly generate and manufacture highly potent antiviral compounds. Our current experience should teach us that we should fully expect future viral outbreaks, and we should not make the same mistake of complacency once we emerge from the current pandemic.

A shining moment for biopharma

Getting back to our normal way of life will be a multistage process that will require solutions from the biopharmaceutical industry. Initially, we'll need to repurpose existing drugs to get us to a stage where we have effective, novel antivirals available, which can then bridge us to when we ultimately have effective vaccines. Since COVID-19 was first detected, the global biopharmaceutical industry has stepped forward to quickly study the utility of currently available drugs, and to aggressively develop novel antivirals and vaccines directed at SARS-CoV-2. We've demonstrated a clear sense of urgency and a willingness to collaborate with other companies, regulatory authorities, governments, and NGOs to accelerate timelines and to find solutions. We've shown the world what this industry is all about, and the critical role it plays in our society.

Our industry attracts some of the world's best, brightest, and most dedicated scientists, clinicians, and entrepreneurs, motivated by a desire to make a positive impact on the world. I truly believe this is a moment where the biopharmaceutical industry, our science, and innovation, will shine, and I'm proud to be a part of this moment in history.

Since 2002, **Dr. John Maraganore** has served as the CEO and a director of Alnylam. Previously, he served as an officer at Millennium Pharmaceuticals Inc. and director of molecular biology, market, and business development at Biogen Inc. At Biogen, he invented ANGIOMAX® (bivalirudin) for injection. He was also a scientist at ZymoGenetics Inc. and the Upjohn Company. He serves on the Agios Pharmaceuticals and BIO Boards.

Alnylam has led the translation of RNA interference (RNAi) into a whole new class of innovative medicines with the potential to transform the lives of patients who have limited or inadequate treatment options. Based on Nobel Prize-winning science, RNAi therapeutics represent a powerful, clinically validated approach. Alnylam was founded in 2002 on a bold vision to turn scientific possibility into reality, which has resulted so far in two marketed products, its robust discovery platform, and deep pipeline of investigational medicines, including six programs in late-stage clinical development.

TOGETHER

Rachel King

I never thought I'd be running GlycoMimetics from my basement, but here I am. Meeting by video with my senior team, talking with investors on our quarterly earnings call, leading company meetings, taking board calls—all from the room that once was an extra guest room, more recently my favorite spot for hosting my book club. Now it's what we jokingly call our new "basement C-suite." I feel less organized here than at the office and less physically comfortable in my work space, and I really miss being able to just walk around and talk informally with my colleagues. But it is working. In the context of all that is happening around the country and the world, this new arrangement hardly qualifies as a sacrifice.

At the beginning of the COVID crisis, I think few people would have predicted the tremendous disruptions we have seen and the length of time we'd be working from home—I know I didn't.

At GlycoMimetics, the initial decision to work from home didn't affect all our colleagues in the same way. As at other companies, many of our office-based teams had already worked from home occasionally, so they had the experience of doing that effectively. For our lab-based teams, however, much of their work simply can't be done at home, and they were frustrated that they could not continue their important projects. They wanted to find a way to continue, while doing what was needed to ensure safety. So, in the beginning, we decided that they could keep going to work in the labs, but in small teams practicing social distancing, which allowed them to continue to progress on our basic scientific programs. Eventually, as conditions worsened and public health required greater protections, our lab-based colleagues went home as well, and have largely been redeployed now to work on other aspects of our programs, though they're eager to get back to the labs. For our office based colleagues, as I write this it has now been nine long weeks at home, and I have been impressed with our productivity. It's not optimal, but we are managing to continue our company operations. I understand that many of our colleagues have a lot of challenges in the current circumstances, and I'm grateful to everyone for how hard they are trying to make this work.

Our main clinical focus for some time has been on our Phase 3 program for our drug candidate uproleselan, which is being tested to treat patients with acute myeloid leukemia, known as AML. That continues to be our focus during the pandemic. Like all clinical trials now, enrollment has dramatically slowed, but, sadly, AML itself has not slowed due to COVID. Patients are still being diagnosed with AML and when that happens, they need to be treated without delay. Our team is working hard to make uproleselan available to them through the trial, to support our clinical sites, and to ensure the integrity of the trial data. I do believe that as hospitals become less stressed by the requirements of treating COVID patients, we'll be able rapidly to get back to increased enrollment of patients in our study. We are all as committed as ever to trying to help people facing this terrible disease.

Everyone is feeling stress and concern. I know these times are particularly difficult for people caring for young children or elderly family members, and for the many people who are now unemployed. There is much to juggle in any given day, and a lot of understandable worries. I hope that in the midst of this, our company can provide some stability and a sense of community for all of us who work together. In addition, we are fortunate that purposeful work is at the heart of what we do at GlycoMimetics. In a health care crisis like this, the tremendous opportunity we have in biotechnology to improve people's health and well-being truly stands out. I hope that for the community of people who make up GlycoMimetics, this work provides a sense of meaning that helps to carry us through these difficult days. Thankfully, the opportunity to have a positive impact on people's lives has always been at the core of what we do, and the pandemic hasn't changed that.

I feel the stress, too, but there is a memory from these COVID days that will stay with me and that gives me strength. It happened the other day in my kitchen, just after my husband, John, and I had had lunch together—one benefit of all this working from home. We had just been standing there talking about our two children, now grown, and about how they were trying to navigate the pandemic. Our daughter and her husband live in North Carolina, where she is an elementary school teacher. She's pregnant with our first grandchild. We talked also about our son and his wife, living in Chicago, where he is a doctor at a big city hospital. We talked, too, about our elderly parents and the concerns we have about ensuring they have all they need, while hoping they stay safe and healthy. They were all on our minds, these people we care so very much about, and who we are now isolated from. Then on the radio came the news that Maryland, where we live, had continued to see a climb in the number of cases and in the number of deaths from COVID. We looked at each other and felt the heaviness of our concerns for the people we love and the latest frightening news. John and I turned toward each other and embraced. We stood there in the kitchen, just stood there for a minute or two, holding each other in silence.

I know we will come through this crisis. In our companies and in our families, we'll do it by relying on and supporting each other—it's what has carried us through challenges in the past, and it's the only true path forward.

Rachel King has spent nearly all her career in the biotechnology industry, with stints at small companies, big pharma, and venture capital. She is currently CEO of GlycoMimetics, a publicly traded company, which she cofounded.

GlycoMimetics is a clinical-stage biotechnology company that uses a specialized carbohydrate-based chemistry platform to discover and develop novel drugs to address unmet medical needs. The company's most advanced drug candidate, uproleselan, is in clinical trials treating patients with acute myeloid leukemia, or AML.

ADAPTING AND INNOVATING THROUGH COVID-19

Richard Pops

Our team at Alkermes is experienced and well-acquainted with complexity. This helped in our response to COVID-19.

We make medicines to treat serious mental illness and addiction, conditions that are often stigmatized and marginalized in our society. What we do is hard, across multiple dimensions. We work to help address complicated public health challenges, interacting with fragmented treatment systems that have entrenched practices and behaviors. Our medicines involve complex dosage forms that require specialized manufacturing capabilities and expertise. Our therapeutic areas are governed by restrictive access and reimbursement policies, comparatively low prices, and higher rebates. With multiple approved medicines and an active development pipeline, we are accustomed to seeing serious problems arise suddenly and developing our own playbook in response.

The emergence of COVID-19 presents a new series of challenges. The pandemic directly threatens the fundamental health and safety of our colleagues, families, and friends in undiscriminating fashion. It changes the way we live and how we protect our families. Schools are dismissed, businesses are shut down, and social distancing is imposed across the country. It affects us all on a visceral level. This amplifies the importance of getting our response right, so we can take care of one another and preserve our ability to meet our public health responsibilities.

Our preparations for COVID-19 began in late January, as we watched the news develop in Wuhan and followed the epidemiologic reports and projections. Whether it materialized into a pandemic or not, we were not going to be caught flat-footed. We formed a senior leadership Crisis Management Team and subteams focused on employee safety and security, internal and external communications, and business continuity to begin scenario planning. By the time the virus hit the US and we learned of local cases, we were ready to take decisive actions.

Our response came in three distinct phases. The first two were planned. The third evolved as we adapted to the new reality.

First came the crisis management phase, when concerns about the danger of the virus were paramount. In this phase, we quickly organized to change the way we do business to protect our employees in the workplace and at home. We had to do this while staying true to a company culture we cherish—one based on collaboration, trust, and a compelling desire to help others. We implemented work-from-home protocols and began to rely on WebEx and Skype. We pulled our field-based personnel and hospital-based sales representatives from in-person interactions with health care providers. We also implemented new protocols to protect the health and safety of our dedicated employees working in our manufacturing facilities and running critical experiments in our labs.

During this phase, communication at every level of the organization took on more importance than ever before. The Crisis Management Team and its subteams communicated regularly to the entire workforce on a range of business and COVID-related topics. There were not enough hours in the day to maintain the connectivity we took for granted when we were all under the same roof. Work and life blended in dining rooms and living rooms and spare bedrooms, as our colleagues balanced shifting responsibilities across all aspects of their lives.

Early on, we held a virtual Town Hall for the entire company. Over 1,700 of our employees joined through their laptops, tablets, mobile phones, and landlines. People were ravenous for updates and connection with their colleagues, and it became increasingly clear that the leaders of the company were becoming a trusted source of unbiased information, assurance, and guidance to navigate these troubling times.

At the Town Hall, I noted a quote from a story published that morning by *Axios* that captured the mood of the country at the moment. It said in part,

". . . the effects of school closures, business restrictions, social distancing and the overload on the medical system are only beginning to set in . . . the findings reflect something between a panic and a 'national malaise,' manifesting in anxiety and uncertainty but also psychological dissonance' . . ."

It was important to acknowledge that people across the country were being deeply affected by this pandemic, so if our employees were feeling that too, it was to be expected. We made it clear that our first priority was their physical and mental well-being. It was essential for everyone to understand that—in effect, nothing else mattered until we got that right. We could not achieve our many goals and progress to the second phase of our response, unless we and our families felt and were indeed safe and secure.

In this second phase, we focused on protecting our business and advancing our 2020 operating objectives. Meeting our operating plan meant far more than revenues and expenses; it included continuing to develop, manufacture, and get our medicines to patients who depend on them. More than 150,000 people received our medicines last year: 150,000 lives affected by alcohol dependence, opioid dependence, or schizophrenia who rely on us to continue manufacturing and providing our medicines. This was our responsibility and our opportunity.

At Alkermes, we are focused on some of the most critical long-term health implications of the pandemic: serious mental illness and addiction. These conditions are exacerbated in the very conditions COVID-19 was driving: fear, social isolation, and economic hardship. During the pandemic and in its wake, identifying and treating patients with these conditions take on a new level of importance.

Across the organization, we had to adapt the conduct of our business in order to maintain continuity. We pivoted quickly to support the needs of our customers and transitioned our customer engagement strategy to a virtual model while advancing our digital capabilities and continuing to support broad access to our medicines. We interacted closely with our clinical trial sites and developed new approaches to collecting data. We evolved our practices in our manufacturing sites and maintained our production schedules. Through all this, we witnessed remarkable innovation and adaptation as smart, committed people refused to let the new reality interfere with their plans.

Along the way, it became clear that new and improved ways of working were going to emerge. Some folks were hunkered down, waiting for it all to end and life as it had been to resume. That was understandable. But others were thinking creatively how to solve problems arising within the context of the crisis and, beyond that, how to catalyze new and more effective practices over the long term. That was exciting.

It is inevitable—some individuals and companies will adapt better than others and emerge from this even more competitive.

I said in a communication to the company:

"I encourage you to be one of the people who finds opportunity in the uncertainty. This moment provides a chance to distinguish yourself and this company from the competition. I cannot believe that we are not going to be more agile and creative, and more intensely focused on doing the right thing for patients and caregivers than others . . ."

So this is the third phase of our response. We are finding ways to learn from these challenging circumstances—to identify new streamlined ways of working, to eliminate unnecessary expenditures, to adapt to the evolving needs of our patient and provider communities, to gain a competitive advantage by being more responsive and creative in meeting the needs of patients, health care providers, and payers.

One example of how health care in our nation could dramatically shift in a post-COVID world may lie in telemedicine, particularly for the ongoing treatment of substance use disorders or serious mental health diagnoses. During the crisis, telemedicine became the only way many patients could interact with their health care providers. Alkermes's medicines—administered by injection once a month or even once every two months—take on new relevance in an environment with less frequent personal interactions with health care providers. In the aftermath of the pandemic, there is a real possibility that accessing doctors virtually becomes much more of a standard of care for certain patients, particularly those with limited mobility and in rural counties.

Health care for these patients could end up being better than it was before this pandemic. We have an opportunity to be leaders in conceptualizing this change and driving policy changes to support increased access to it.

Another example is how we have adapted to seek and listen to the needs of the communities central to our mission and deeply affected by the pandemic. When COVID-19 first began shutting down and threatening the health of our country, we reached out to our patient communities and their advocates, asking about the new challenges patients faced and what might be done to address them. We learned of a broad set of needs that continue to evolve. We realized we need to move quickly and act, to connect communities with needed tools and support that go beyond our medicines.

In early May 2020, we announced the launch of the Alkermes COVID-19 Relief Fund, to address COVID-19-related needs for people living with addiction, serious mental illness, or a cancer diagnosis. We will support programs that can be implemented within a short time and have the potential to lead to sustained impact beyond the immediate crisis.

As I write this, all three phases of our response are ongoing. We are still in the midst of the pandemic, uncertain about the timing and the extent of any return to normalcy. What is remarkable is that, despite the constraints and limitations, we are operating, advancing, and evolving as we provide essential medicines to people with serious conditions.

We expect to come out of this even more focused and effective than we were going in.

Our work will blend into a broader mosaic of brilliant work being done by the biopharmaceutical industry. Science, medicine, and compassion will solve this problem, and our industry will be at the core of the solution. At a moment in time when policymakers and the public are skeptical about the values and virtue of our industry, we have a clear opportunity to demonstrate them for the benefit of the entire world.

Richard Pops is the chair and CEO of Alkermes. He is member of the board of directors of Neurocrine Biosciences and Epizyme, both publicly traded biopharmaceutical companies, and also BIO and PhRMA, the industry's principal advocacy organizations.

Alkermes is a fully integrated, global biopharmaceutical company developing innovative medicines in neuroscience and oncology. The company has a portfolio of proprietary commercial products focused on addiction and schizophrenia, and a pipeline of product candidates in development for schizophrenia, bipolar I disorder, neurodegenerative disorders, and cancer.

LEADING THROUGH UNPRECEDENTED TIMES WITH RESILIENCE AND CLARITY OF PURPOSE

Deborah Dunsire

On January 11, as biotech and pharma world leaders poured into San Francisco to start the annual J.P. Morgan Healthcare Conference, the first death in China from COVID-19 was reported. I remember thinking that this might be another SARS challenge for China to manage. Meanwhile, life and business for all of us outside China went on at the usual hectic pace, with a focus on driving the business year forward.

As January progressed, we watched in disbelief as Wuhan was locked down and we saw our Lundbeck China team sidelined. Working from home became their new normal. On February 4, we met with our Chinese leadership team by videoconference and decided to model an anticipated impact on our business in China. But we still had no real insight about the global threat growing underneath us. In the last days of February, the first Danes were confirmed to have COVID-19, many having returned from ski vacations in Northern Italy. February 28 saw the first reported US death, in Washington State. The specter of a global pandemic dawned in our thinking, but even then we could not imagine that within two weeks the country would be on lockdown. During March, we saw the unthinkable happening around the world: a surge in cases across continents, overwhelmed health systems in Wuhan, and then Northern Italy, and so on. We collectively experienced the staggering impact of the unrelenting upward march in the death toll.

As company leaders, we were thrust into an unimaginably VUCA (volatile, uncertain, complex, ambiguous) world. We had no roadmaps and no sure way to assess the future risk. Our first focus was to ensure that our employees were safe, and we strenuously advised them to heed the call for nonessential workers to shelter in place and work from home, as they already were in Italy. Lundbeck is dedicated to restoring brain health, so every person can be their best. That guiding star had us focus immediately on how to preserve the supply of medicines that are a lifeline for people facing mental illness. In our four production units, in Denmark, Northern Italy, and Southern France, our teams were clear-sighted and resolute—crafting plans to streamline teams to allow for appropriate physical distancing in factories and labs as they continued working. Their courage and creativity have been a beacon for us all.

For all our teams around the world, thrust into the existential crisis and the isolation, we knew we had to focus on building resilience to ensure that they could manage their own stress and fear, all while inventing new ways to support our patients and the health care professionals who care for them. Over the weeks, the leadership team invested a significant effort in communicating with our people and helping them prioritize actions to help make them resilient to the rising stresses. Some ideas we shared were:

- Focus on purpose, why we do what we do. Brain health is even more vital in these times.
- Acknowledge that the current situation is stressful. Act to mitigate its impact with good self-care.

- Look actively for opportunities brought on by the challenge—living and working differently.
- Remind yourself and others that we will get through this.
- Envision what is needed to be advance in the "new normal."
- Practice optimism! Actively reflect on the good things that are present.
- Take positive action—even small steps forward bring a sense of accomplishment.
- Keep community—we are stronger together.
- REST!

I have been so impressed with how our colleagues have come through this crisis to date and found new ways forward. The reality of our "new normal" will call for an even deeper resilience, as the world will not be able to return to the way it was for many months. As this VUCA reality evolves, we know that it will bring with it a tsunami of another sort. We are witnessing the emergence of a mental health crisis of unprecedented proportions. We don't yet fully understand the toll of the pandemic, but we recognize there will be significant, long-lasting, and heartrending mental health reverberations.

Today we are seeing dramatic spikes in rates of anxiety and depression among those who did not previously have a diagnosed mental health condition, along with worsening of symptoms and relapse in people already diagnosed with serious mental illnesses like bipolar disorder, major depressive disorder, and schizophrenia. Rates of post-traumatic stress disorder are soaring among first responders, health care workers, and others directly impacted by loss or illness. This mental health emergency threatens global health. We know that people with mental health conditions are at higher risk of chronic physical health conditions like heart disease, diabetes, and Alzheimer's disease. The tragedy of loss of hope that has resulted in suicide is rising alarmingly.

It is becoming abundantly clear that we must address the mental health fallout from the pandemic before it reaches unfathomable proportions. This will require the continued resilience of companies like Lundbeck, with our unwavering commitment to brain health, and the collective resilience of our global community to finally prioritize effective mental health care. As the only global company solely focused on supporting people impacted by brain diseases, Lundbeck is well-prepared to lead through this challenge. We have a long commitment to reducing the stigma of mental illness, eliminating barriers to care, and advancing new and better treatments for people living with mental health disorders. We are uniquely positioned to lend our expertise and energy to flattening the mental illness curve—and we already are actively engaged in doing so.

But we know that to truly flatten the mental illness curve, a broader response is required. While COVID-19 is exacerbating the incidence of mental ill health, this crisis has been simmering just below the surface for years. Pre-pandemic, about one in ten people worldwide were living with a mental disorder, with many more likely undiagnosed and many lacking access to care and appropriate treatment. Over the past decade, we saw an increase in the prevalence of mental health conditions, particularly among young people. Suicide is now the second-leading cause of death among young adults between the ages of fifteen and twenty-nine.

Across all ages, mental disorders are among the leading causes of ill health and disability worldwide. The World Economic Forum estimates that by 2030, the global cost of mental illness will rise to $6 trillion, more than double the total cost of cancer, diabetes, and cardiovascular diseases combined—and that's a prepandemic estimate.

As a society, we cannot afford to ignore the yawning gaps in our mental health care prevention and treatment around the world. Lundbeck is firmly committed to contributing our own efforts to bending the looming mental illness curve, while also advancing a multisectoral approach to long-term mental health promotion. We are a signatory to the UN Global Compact and are committed to working alongside public and private partners to advance the UN Sustainable Development Goals, which include promotion of mental health and reducing deaths from suicide. We are working alongside policy makers, physicians, patients, and communities around the world to raise awareness of the burden of mental illness, drive parity of care, implement universal screening and treatment for mental health, strengthen capacity of the behavioral health workforce, and more. All that is required to return to resilience as a society.

And, of course, our focus on innovation in brain science must and will endure. While our expertise does not lie in developing vaccines or treatments for the novel coronavirus, we are focusing on meeting this watershed moment for mental health. We are harnessing our expertise across the enterprise to rethink how we and others can tackle these entrenched issues differently.

The work of the coming months (and likely years) will require that we—Lundbeck's nearly 6,000 dedicated colleagues across the globe—tap our own reserves of resilience. For resilience is not only about the capacity to recover quickly after setbacks; it also is about preserving clarity amid chaos. It is there, in clarity of our purpose—tirelessly dedicated to restoring brain health—that we are determined to meet the second wave and bring out the best in each of us for the benefit of global brain health.

Dr. Deborah Dunsire is president and CEO of H. Lundbeck A/S, the only global biopharmaceutical company singularly focused on restoring brain health so every person can be their best. A physician trained in South Africa, she has over three decades of leadership in biotech, from large companies to start-ups. She is on the boards of Alexion and Ultragenyx, and was on the boards of Takeda and Allergan.

BROAD LESSONS

BIOETHICS AND THE NEED FOR ETHICAL LEADERSHIP DURING THE COVID-19 PANDEMIC

Kenneth I. Moch

Throughout history, medical crises have accelerated the development of new medical products and procedures.[1] The COVID-19 pandemic is such a crisis, threatening human life in a potentially uncontrollable and unpredictable manner.

The biotechnology community has responded to this pandemic by marshaling its enormous skills and scientific resources toward developing new medical countermeasures—particularly therapeutics and vaccines. This has been done with such intensity that as of this writing over 950 inquiries for drug and vaccine proposals have been received by the FDA ("an overwhelming amount in a short period of time"),

1. Barr J, Podolsky SH. A National Medical Response to Crisis—The Legacy of World War II. New England Journal of Medicine. April 29, 2020. doi: 10.1056/NEJMp2008512.

over seventy clinical trials are ongoing in the US, and over 200 other development programs are in planning stages.[2]

Beyond evaluating the benefits and risks of each potential medical countermeasure to treat the infected, control the spread, or prevent recurrence, many fundamental ethical questions must also be addressed. The most complex are these: Who decides the specific parameters of each clinical development pathway; how are the decisions made; and what are the near- and long-term impacts and implications of these decisions? Said differently, who decides "who shall live" and adjudicates the ethical implications of decisions that are inherent in the clinical development pathway for new medical countermeasures?

Philosophers and bioethicists often look at questions like these through the lens of the "Trolley Dilemma," which revolves around the ethical and moral complexities of deciding who to save in a life-or-death situation and whether such a decision passively or actively causes harm to others.[3] In the face of the COVID-19 crisis, biotechnology companies are developing a spectrum of new medical countermeasures under extreme time and resource constraints coupled with an incomplete understanding of the course and the complications of this new disease. As a result, biotech company leaders are facing extraordinarily complex ethical questions akin to the Trolley Dilemma. Among the questions are these:

- How do you evaluate benefits and risks to different patient populations at different stages of the disease? What are the impacts of these risk/benefit decisions on the applicability of the resultant data to current or future patients in those other population segments?
  - Where in the disease progression is an experimental medical countermeasure best developed? For example, should the target

---

2. Woodcock J. Current State of COVID-19 Treatment Trials: US Perspective. Presentation to Clinical Trials Transformation Initiative meeting. April 23, 2020. https://www.ctti-clinicaltrials.org/briefing-room/webinars/designing-high-quality-covid-19-treatment-trials.

3. Hacker-Wright J. "Philippa Foot". The *Stanford Encyclopedia of Philosophy* (Fall 2019 Edition), edited by Zalta EN. https://plato.stanford.edu/archives/fall2019/entries/philippa-foot.

  patient population include only critically ill patients, those recently infected, or individuals who have never been exposed?
  - At what stage in the clinical or preclinical development process can a medical countermeasure be administered to humans?[4] For example, is it acceptable for critically ill patients to be given compounds not yet tested in humans? What about individuals who were recently infected, or individuals who have never been exposed?
  - What inclusion and exclusion criteria should be established for clinical testing? For example, in addition to disease status and patient prognosis, what are the criteria for comorbid conditions, concomitant medications, age, or gender? How do companies balance the demands of social justice, developing interventions that are useful for the broadest patient populations, with the time and resource constraints and the overarching societal demand for rapid clinical development?
- Can clinical trials be randomized and placebo-controlled when the standard of care might lead to death, or should they be open-label? How does this affect the evaluation of potential endpoints and the decisions regarding clinical trial event rates for these endpoints?
  - As the number of clinical trials increases, will it become difficult to attain the targeted enrollment number as each trial competes for patients, thus unintentionally delaying the time it takes to clinically prove the potential of new medicines? How should competing trials be prioritized, and if so in accordance with what principles?

Effective ethical leadership requires grappling with these difficult questions. It is the responsibility of the leadership of each biotechnology company to make the specific decisions regarding the preclinical

4. U.S. Food and Drug Administration. Public Workshop—Clinical Trial Designs for Emerging Infectious Diseases. November 9-10, 2015. http://wayback.archive-it.org/7993/20170111002901/http:/www.fda.gov/EmergencyPreparedness/Counterterrorism/MedicalCountermeasures/AboutMCMi/ucm466153.htm.

and clinical development pathway for their experimental medical countermeasures as well as to evaluate the ethical implications of such development plans.[5] Not asking these questions and not taking an active position is a decision itself.

Science must guide the process. Decisions must be based on the clinical data generated at each decision point coupled with the collective wisdom and experiences of the scientists, clinical development professionals, and regulators. And there is nothing wrong with providing thoughtfully reasoned hope and comfort to a worried population[6], such as talking about the potential for developing new drugs and vaccines.

But speculation in the absence of data can be deadly. Decisions cannot be made or influenced by those who use hyperbole or promote false hope. External pressures and misstatements from private individuals (e.g. "antivaxxers"), members of the media, or government officials who do not clearly understand the complexities of and timelines for drug and vaccine development can only cause the misallocation of resources, create false hope and inappropriate actions, and harm those in true need of help.[7] That this has happened in the United States during the COVID-19 pandemic, coupled with delays in taking action despite significant signals and a national pandemic preparedness strategy[8] that was apparently disregarded, will be the source of many future analyses.

---

5. Moch KI. Wanted: Guidelines for Access to Experimental Drugs. Wall Street Journal. March 16, 2015. https://www.wsj.com/articles/kenneth-i-moch-wanted-guidelines-for-access-to-experimental-drugs-1426547602.

6. Zagury-Orly I, Schwartzstein RM. COVID-19—A Reminder to Reason. New England Journal of Medicine. April 28, 2020. doi: 10.1056/NEJMp2009405.

7. Caplan AL, Moch KI. Speculation in the absence of data can be deadly. New York Daily News. April 28, 2020. https://www.nydailynews.com/opinion/ny-oped-speculation-in-the-absence-of-data-can-be-deadly-20200428-tsh6y7wh3fbr7a5lzftz6cvkte-story.html.

8. U.S Department of Health and Human Services, Office of the Assistant Secretary for Preparedness and Response. 2017-2018 Public Health Emergency Medical Countermeasures Enterprise (PHEMCE) Strategy and Implementation Plan. December 2017. https://www.phe.gov/Preparedness/mcm/phemce/Documents/2017-phemce-sip.pdf. See also

This is not to say that clinical development norms are immutable in the face of a crisis. Many countries and their regulatory agencies, as well as hospitals and medical professionals, have actively or tacitly altered their view of risks and benefits.[9] This gives rise to another key ethical question:

- How sacrosanct is the clinical development process[10], and what alterations can or should be made in the pursuit of shorter development timelines for individuals facing life-threatening conditions who have no other therapeutic options?

The past several decades have seen the rise of preapproval access (PAA) to experimental medicines, commonly referred to as compassionate use or expanded access, under which select critically or terminally ill patients receive unapproved medicines for treatment purposes.[11,12] PAA treatment protocols for individuals or groups of patients are open-label by design, and generally collect significantly less data that can be used to help determine the potential of the experimental medicine than are collected in clinical trials. The debate

Sanger DE et. al. Before Virus Outbreak, a Cascade of Warnings Went Unheeded. The New York Times. March 19, 2020. https://www.nytimes.com/2020/03/19/us/politics/trump-coronavirus-outbreak.html.

9. Cook D. et al. Clinical research ethics for critically ill patients: A pandemic proposal. Critical Care Medicine. April 2010. doi: 10.1097/CCM.0b013e3181cbaff4.

10. Bothwell LE, Podolsky SH. The Emergence of the Randomized, Controlled Trial. New England Journal of Medicine. August 11, 2016. doi: 10.1056/NEJMp1604635.

11. Speers MA. Providing Patients with Critical or Life-Threatening Illnesses Access to Experimental Drug Therapy: A Guide to Clinical Trials and the US FDA Expanded Access Program. Pharmaceutical Medicine. March 19, 2019. 33:89–98. https://doi.org/10.1007/s40290-019-00274-3.

12. Moch KI. Ethical Crossroads: Expanded Access, Patient Advocacy, and the #SaveJosh Social Media Campaign. Medicine Access @ Point of Care. Volume 1, January–December 2017. https://doi.org/10.5301/maapoc.0000019.

about PAA during a medical crisis is not new[13]; for example, during the 2014 Ebola outbreak the use of experimental medicines that had not previously undergone any human testing but had shown promising preclinical results was considered ethically acceptable under strictly specified criteria.[14]

It is possible and perhaps likely that competition for patients between clinical trials will hamper each trial's ability to attain the targeted enrollment, and thus delay the timelines. While the COVID-19 pandemic is larger and more geographically dispersed than the Ebola epidemic, the number of potential medical countermeasures being evaluated is dramatically larger. The inability to complete trials focused on a more localized epidemic has happened in the past, albeit under conditions relating to declining potential patient populations for testing.[15,16]

Concern about identifying sufficient numbers of individuals to enroll in COVID-19 clinical trials have already been raised within the context of vaccine development, with the concept that volunteers could consent to being dosed ("challenged") with live COVID-19 virus.[17] Highly controlled "human challenge studies" could be designed as alternatives to large field efficacy trials, with the multiple goals of im-

13. Richardson, T., Johnston, A. M., Draper, H. A Systematic Review of Ebola Treatment Trials to Assess the Extent to Which They Adhere to Ethical Guidelines. PloS one. January 17, 2017. https://doi.org/10.1371/journal.pone.0168975.

14. World Health Organization. Ethical Considerations for Use of Unregistered Interventions for Ebola Virus Disease (EVD): Report of an Advisory Panel to WHO. 2014. https://apps.who.int/iris/bitstream/handle/10665/130997/WHO_HIS_KER_GHE_14.1_eng.pdf.

15. Chimerix Scraps Testing of Experimental Ebola Drug in Liberia. Wall Street Journal. February 1, 2015. https://www.wsj.com/articles/chimerix-scraps-testing-of-experimental-ebola-drug-in-liberia-1422739087?mod=article_inline.

16. Gilead. A Trial of Remdesivir in Adults With Severe COVID-19. Clinicaltrials.gov. Posted February 6, 2020. ClinicalTrials.gov Identifier: NCT04257656.

17. Plotkin SA, Caplan AL. Extraordinary Diseases Require Extraordinary Solutions. Vaccine. May 19, 2020. https://doi.org/10.1016/j.vaccine.2020.04.039.

proving the understanding of the course of the disease, comparing efficacy of and between multiple vaccines, and accelerating vaccine licensure.[18,19] Clinical trial proposals like this give rise to additional ethical questions that are more appropriately answered at the societal level, as they represent overarching ethical issues:

- What level of risk should asymptomatic participants accept (or be allowed to accept) to help develop treatments for themselves or for others who are currently infected or might be in the future?
- Given different societal norms in different countries, will international regulators collaborate and share information to facilitate efficient approvals? And if not, can or should trial designs be altered, and what are the implications of such changes?
- Should some patient populations, e.g. the elderly, be asked or be willing to take higher risks to help accelerate the availability of new medicines? Should these same individuals be exposed to greater risks for the sake of the economy?

A final ethical dilemma that the leadership of the biotechnology industry must address relates to the needs of patients who now or in the future will face a life-threatening condition other than COVID-19. The intense and necessary emphasis on developing COVID-19 medical countermeasures has drawn significant resources away from the development of new therapies for other life-threatening diseases. Particularly for highly prevalent chronic diseases which have long development timelines and immense development costs, such as diabetes, Alzheimer's and cardiovascular disease, the delay in developing new medicines and thus the impact on these patient populations is

18. World Health Organization. Key criteria for the ethical acceptability of COVID-19 human challenge studies. May 6, 2020. https://apps.who.int/iris/bitstream/handle/10665/331976/WHO-2019-nCoV-Ethics_criteria-2020.1-eng.pdf?ua=.

19. Eyal N, Lipsitch M, Smith P. Human Challenge Studies to Accelerate Coronavirus Vaccine Licensure. *The Journal of Infectious Diseases*. March 31, 2020. https://doi.org/10.1093/infdis/jiaa152.

likely to be substantial. The decision as to when and how to redirect resources back to these other diseases will have significant implications for future individuals at their time of need.

There are no simple, monolithic answers to any of these ethical questions, if there are answers at all. More often, leaders will have to make "judgment calls" based on the currently available scientific information and their collective experiences and wisdom. No matter what, these decisions need to be actively made, with due consideration to the ethical implications for "who shall live."

Kenneth I. Moch has been cofounder or CEO of five companies that pioneered novel therapies for life-threatening diseases, resulting in multiple marketed products. He chairs the Bioethics Committee of the Biotechnology Innovation Organization, is a founding member of the NYU Working Group on Compassionate Use & Preapproval Access, and has authored or coauthored numerous articles on medical ethics and preapproval access to experimental medicines.

During the COVID-19 pandemic, Ken is serving as Senior Advisor to the Chairman of the Center for Global Health Innovation and the Global Health Crisis Coordination Center.

CORONAVIRUS AND PEOPLE OF COLOR: THE MAGNIFICATION OF HEALTH DISPARITIES AND WHAT WE MUST DO NOW

Quita Highsmith

For many of us, the coronavirus pandemic has provided more time for reflection on what is truly important. For me, as a Black woman and biotech industry leader, I'm particularly concerned about how the COVID-19 public health crisis is amplifying health disparities that have harmed communities of color for decades.

Due to a variety of wide-ranging inequities, people of color are less likely to have steady access to quality medical care.[1] They are less likely to live in neighborhoods with healthy food options[2] and more

1. https://www.ncbi.nlm.nih.gov/books/NBK24693/

2. https://www.ucsusa.org/resources/devastating-consequences-unequal-food-access

likely to face exposure to environmental factors that impact their health. People of color are also more likely to work in the service industry or the frontline health care field, or have lower-wage jobs that do not offer paid sick leave. During the COVID-19 pandemic, we are seeing a disproportionate number of people from historically underserved communities experience higher infection risks,[3] with increased hospitalization and mortality rates. Preliminary data from New York City is showing that COVID-19 is killing Black and Hispanic people at twice the rate of white people,[4] and similar trends are being reported for cities in Louisiana, Michigan, and Illinois.[5]

A lack of access to and participation in clinical research[6] is one systemic barrier hindering optimal health outcomes in historically underrepresented communities today. The imbalance of representation within clinical research was recognized by the federal government almost thirty years ago, with the passing of the NIH Revitalization Act of 1993.[7] This legislation mandated that all federally funded clinical research include women and racial and ethnic minority groups. Since then, women have reached parity, with over 52 percent of participants in NIH-supported clinical research reported in 2018, but for members of racial minority groups, this number was only 29 percent, and for ethnic minorities, 9 percent.[8]

We can no longer accept this status quo. To break this cycle of inequity and create a world where all patients can access investigational medicines, we must come together as industry leaders to take bold and decisive action.

---

3. https://www.cdc.gov/coronavirus/2019-ncov/need-extra-precautions/racial-ethnic-minorities.html

4. https://www1.nyc.gov/assets/doh/downloads/pdf/imm/covid-19-deaths-race-ethnicity-04162020-1.pdf

5. https://www.newyorker.com/news/our-columnists/the-black-plague

6. https://www.theatlantic.com/health/archive/2016/06/why-are-health-studies-so-white/487046/

7. https://history.nih.gov/research/downloads/pl103-43.pdf

8. https://nexus.od.nih.gov/all/2019/05/06/nih-inclusion-data-by-research-and-disease-category-now-available/

Building the foundation

In my previous role as head of alliance and advocacy relations at Genentech, I saw firsthand how difficult it was to find diverse patients who had participated in clinical trials. I learned that this gap extends beyond diversity in patient selection into almost every aspect of clinical research—including the diseases we choose to study, where the sites are located, and the investigators themselves. This journey of questioning and exploration led me to cofound Advancing Inclusive Research™ with Nicole Richie, PhD, global head of health equity and population science at Genentech. This initiative is a priority for our company and a key strategic focus of the diversity and inclusion office.

Over the past several years, we have changed how we approach our clinical research programs in these ways:

- In our earliest communications with site investigators, we emphasize our commitment to addressing barriers to participating in clinical research.
- We modified our contract language to encourage inclusion of underrepresented patients.
- We changed our inclusion/exclusion protocols to be more inclusive of patients with diverse ethnicity (e.g. race- or sex-estimated glomerular filtration rate vs. absolute creatinine cutoff, benign neutropenia in certain ethnic groups or elevated serum creatinine in patients of African descent).

Additionally, to gain much-needed insight from experts, we formed an external council composed of physician thought leaders, academic research experts, and patient advocates. This collaboration has provided the opportunity to raise awareness with a broader audience and cocreate solutions to address inequity in clinical research.

Equally important is a continued partnership with government agencies, such as the Office of Minority Health and Health Equity at the Food and Drug Administration (FDA), with whom we share a similar vision and regularly dialog about relevant programs and initiatives.

As a result of these efforts, we are now positioned to deliver innovative new studies, such as a first-of-its-kind clinical trial, CHaracterization of ocrelizumab In Minorities with multiplE Sclerosis (CHIMES, Reference ID# ML42071), to study a treatment for multiple sclerosis specifically within the Black and Latinx patient populations.

Taking action during a pandemic

The emergence of COVID-19, and the disparities we see in incidence and mortality rates, has accentuated why it is so critically important to expand this work. Genentech is actively including historically underrepresented groups in its COVID-19 clinical research. In collaboration with the Biomedical Advanced Research and Development Authority, Genentech moved quickly to set up a study to evaluate the safety and efficacy of intravenous Actemra® (tocilizumab) in adult hospitalized patients with severe COVID-19 pneumonia (COVACTA, Reference ID# NCT04320615).

In this trial, we have intentionally included sites at the intersection of COVID-19 hot spots and underrepresented populations. For example, one of the first COVACTA patients in the United States came from Ochsner Medical Center in New Orleans, which has a large minority population with high Medicaid enrollment. We had not partnered with Ochsner in infectious disease previously. Naive site selection can be incredibly challenging for companies to enroll patients under tight timelines. However, we knew we needed to prioritize locations with communities of color to take strides toward solving health inequities.

We didn't stop there as early data suggested that the COVID-19 pandemic may be disproportionately affecting underserved and minority populations. We took this early data to hospital administrators and physicians at inner-city hospitals in New York City and our medical affairs organization cocreated another randomized, placebo-controlled study called EMPACTA to evaluate the efficacy and safety of tocilizumab in hospitalized patients with COVID-19 pneumonia. The study will enroll patients with COVID-19 pneumonia at hospitals across the United States that reach underserved communities, including several public hospitals in New York City. These examples show that bold investments are needed to improve the health disparities that exist. This is the deliberate type of scale that is required from the industry.

The path forward

Throughout my career, I've been inspired by the advancements we've achieved in science and medicine. It gives me great hope for the future. What I also know to be true is that clinical research is not benefiting all patients equally, and some communities are being left behind.

We face a great deal of uncertainty in the coming months about COVID-19 and its impact on our communities. In the interim, we cannot wait to start creating a more equitable future for all patients, which includes a steadfast commitment to increasing diversity in clinical research.

In order to create more inclusive clinical trials, we must start with the fundamentals:

- Ask clinical trial site investigators to be intentional about recruiting diverse patients.
- Expand site networks to include potentially smaller, newer sites with more diverse catchment areas.
- Prepare guidance documents with clear expectations for internal teams and external partners.

We can use this moment as a catalyst for action. Genentech's commitment to being a change-maker has never been stronger. Each of us, whether a trial sponsor, a physician, a legislator, a caregiver, a regulator, or a patient, has a role to play.

What we have seen in response to COVID-19 is that, when the entire health care ecosystem comes together with a sense of urgency and purpose, we can overcome obstacles in ways we didn't think possible. Not only are we seeing clinical trials designed and approved faster than ever before, but we're seeing it can be done while including underrepresented populations. Imagine what we can achieve if we bring this same mindset and determination to solving the health disparities crisis for people of color.

Let's be bold together.

Quita Highsmith is the chief diversity officer at Genentech. She is an award-winning visionary with more than three decades of experience and a named author on health disparities in publications including the Journal of Oncology Practice. She leads the diversity and inclusion strategy for Genentech, including developing talent, supporting STEM programs, and addressing disparities via Advancing Inclusive Research.

Founded more than forty years ago, **Genentech** is a leading biotechnology company that discovers, develops, manufactures, and commercializes medicines to treat patients with serious and life-threatening medical conditions. The company, a member of the Roche Group, has headquarters in South San Francisco, California. For additional information about the company, please visit gene.com.

AN URGENT CALL TO ACTION FROM THE RARE DISEASE COMMUNITY

Luke Rosen

On February 28, I stood on a crowded line waiting to pass through the Canadian border checkpoint at Montreal's Trudeau Airport. Ordinarily I would have balked when an agent stopped to spray sanitizer on the hands of every passing traveler. Any other day I probably would have paused with concern when I saw border agents wearing N95 masks. I certainly should have questioned more when the Canadian customs agent asked twice if I had recently traveled to China. Any other day I would have identified the increased precaution people were taking and sensed the palpable panic about to erupt. But on this day, I absorbed nothing—certainly not the possibility that a virus was about to cripple civilization. On February 28, standing on a crowded line in a foreign country, I was numb and consumed with sadness. I was terrified the agent would ask me to remove my sunglasses only to reveal tears in my eyes. I was on my way to Lhassa's funeral and thinking about what I would say to her father; a father who just lost his only daughter. Lhassa was a five-year-old girl whose death was caused by KIF1A Associated Neurological Disorder, the rare disease my daughter was diagnosed with in 2016. This would be the fourth funeral of a KIF1A child I traveled to in as many years.

Like every family in our rare disease community, Lhassa's has become an extension of my own. This community is bonded by a love so clear that even a language barrier could not delay the most painful conversation my job calls for; a conversation I have had seven times since our KIF1A families and doctors joined in a mission to find treatment for children living with this rare neurological disease. Hours before Lhassa died, I spoke with her mother—a conversation far too regular in the past four years, and a conversation that kills part of my soul every time. I explained to my close friend, as she held Lhassa in the final hours of her life, that an organ donation team was on call and ready to make an unimaginable process as seamless as possible. I asked Lhassa's mom to donate her dying daughter's brain and spinal cord to researchers in New York. We talked about how Lhassa would live on in science as an ever-present facilitator for researchers to understand KIF1A. I explained that studying the pathology of her daughter's brain would allow researchers to discover treatment—a treatment that did not come in time for her own little girl. Words on this page will never describe the unfair and horrific trauma that accompanies those unthinkable end-of-life conversations with parents about to lose their child.

On February 28 I was standing on a crowded line at an international airport and thinking about my daughter, Susannah. I was thinking about Lhassa. I was thinking about children dying from rare neurological diseases. And I was crying. I did not notice that everybody around me was preparing for a catastrophic global pandemic.

In later weeks, COVID-19 ravaged the world. The impact on rare disease communities is beyond measure. For rare disease patients and families, the global response to this crisis is paradoxical. On one hand, our rare disease community has been devastated by painful oversight, lack of care, and diminished quality of life. In parallel to this devastating impact on the rare disease community, in this pandemic, hope for faster discovery of cures has been dangled in front of us.

Despite a president whose delay fueled fatality, the rare disease community has witnessed a swift industrial response and remarkable pace of research to develop COVID-19 treatments and diagnostics. The translational speed and medical blitz undertaken by industry to combat COVID-19 is proof that treatments can get to patients much faster. COVID-19 proves that when innovators galvanize with focused resources and urgency, treatments are developed efficiently to save people who are dying of rare diseases.

An abhorrent lack of leadership is responsible for regression in people affected by degenerative diseases like KIF1A. Children with complex medical needs were forgotten as governments closed public schools. Special needs students who rely on services like physical therapy, occupational therapy, speech therapy, and other crucial interventions were sent home to isolation with no emergency action plan. Without consistent services, our children lose words, steps disappear, and skills vanish—skills that parents, therapists, and educators fight relentlessly to preserve. These vital therapies do more than maintain and improve skills; they buy us time while we work to find treatments and cures. Critical medical appointments and medication changes are missed and standard of care has become an improvisation with little direction. If an emergency room visit is needed, lifesaving medical equipment is allocated elsewhere because our children are admitted alone, have cognitive delays, and are unable to communicate. People living with rare neurological diseases are suffering, and their lives have been de-prioritized because of this pandemic.

Foundations and advocacy groups work tirelessly to fund, support and accelerate discovery of treatment for rare diseases. In many instances, advocacy and family funding are primary sources for early scientific discovery. This integral preclinical research has been derailed as government and academic initiatives pivot to COVID-19. According to a webinar presented in April by *Research America*, "This has stopped scientific investigations that are essential to understanding, treating and curing countless diseases. Those studies have just been stopped, abruptly terminated, labs closed."[1] Discovery of rare disease

1. https://www.researchamerica.org/news-events/events/alliance-member-meetings

treatment is an afterthought as academic labs are locked down, animal models euthanized, and precious patient-derived cell lines rendered inaccessible or even destroyed.

Amid this rare disease derailment, industry response to COVID-19 created a window of opportunity, and rare disease champions must act right now. This crisis triggered resources and workforce to spontaneously develop therapeutics for COVID-19. Biotech leaders executed in weeks what the rare disease community has been fighting for years to accomplish: diagnostics and therapeutics.

Families living with rare diseases know time is as much our enemy as nature. Every second together is fleeting, and COVID-19 forced the whole world to feel an identical terror; a terror that rare disease families live with daily. The emergency of COVID-19 induced a unified commitment and work ethic that, if scaled, will bring treatment to rare disease patients before time runs out. The heroism of people on front lines define our society, and previously underappreciated workers are rightfully recognized as the engine that drives progress—and saves countless lives. Now more than ever, together we are strong. A newfound global empathy has emerged for people suffering from catastrophic diseases with no treatments or cures.

Because of this global pandemic, humanity is teeming with science-driven empathy and commitment to every patient and family. My call to action for industry is clear: seize this opportunity. Do not abandon rare disease families. Attack rare disease therapeutics with the same relentless work ethic. Industry must act with continued vigor and utter urgency to accelerate treatments for rare diseases.

Biotech response to COVID-19 proves that the burden to accelerate treatments and cures for rare disease no longer falls solely on shoulders of parents, families, and advocates. Industry response to this pandemic is evidence that we can harness existing science, execute innovation, and bring treatments to rare disease patients exponentially faster. COVID-19 is a model and road map for accelerated rare disease research, development, approval, and access—a pathway of hope forged from a devastating medical crisis. A moment in time that will reveal the character of industry leaders.

Luke Rosen, MSc, is vice president of accelerated development and patient engagement at Ovid Therapeutics. Rosen is founder of KIF1A .ORG, a nonprofit organization started in 2016 to accelerate discovery for KIF1A Associated Neurological Disorder, a rare disease affecting his daughter, Susannah. In 2017, Rosen left his career as an accomplished actor to support research, discovery, and development of treatments for rare, neurological, and genetic diseases.

Science-driven and patient-focused, **Ovid Therapeutics** works to bring treatments to people affected by rare neurological and genetic diseases. Ovid seeks to understand what matters most to families living with rare diseases, and to address the challenges they face every day.

AGRICULTURAL BIOTECHNOLOGY: ENSURING OUR FOOD SUPPLY

Sylvia Wulf

The COVID-19 pandemic has been a catalyst for me to reexamine my purpose and priorities. I am passionate about serving the marginalized, poor, and hurting in this world. I'm passionate about caring for God's creation. It's the reason I accepted the position of CEO of AquaBounty, because I truly believe global acceptance of biotechnology is necessary to feed the world sustainably, and I want to be part of making that happen. As I reflect on the challenges the world currently faces and the impact of the global pandemic, I am now more convinced that using biotechnology and innovation in agriculture will be critical to ensure a safe, sustainable, and secure food supply.

In developed countries, few people worried about going hungry. We grew spoiled by an accessible and affordable food supply. And even though biotechnology positively affected that accessibility and affordability, critics were able to dominate the dialog on genetically modified organisms. AquaBounty's journey is a stark example of the unfair and unscientific hurdles facing both crop and animal biotechnological innovation.

On April 5, 2020, The UN Food and Agriculture Organization alerted the world that a food crisis as a result of the COVID-19 pandemic was a real risk. The report said: "The COVID-19 pandemic is likely to have severe repercussions for vulnerable rural and urban populations. Households could be affected by a decline in their purchasing power, while at the same time facing surging prices for some food items, unavailability of products due to supply chain disruptions and containment policies that could limit access to markets. Such impacts would significantly affect the lives and livelihoods of already vulnerable households dependent on food production and livestock rearing in particular. Despite a tendency to be inelastic, the demand for food is at risk of declining, particularly in poorer countries and for higher-value products. Uncertainty could also increase the likelihood of social tensions and conflict."

In developed countries, only a small percentage of the population is engaged in agricultural production. People no longer understand where their food comes from or how farmers and ranchers work diligently to provide that food. We've rightly heralded the many health care workers on the front lines of the pandemic, but without agricultural workers and others in the food supply chain, most of us would starve. Yet modern society seems to place little value on their efforts and ignores the challenges they face with every cycle of production.

Now is the time to rethink and reposition the importance of scientific innovation, specifically biotechnology's impact on our food supply. The crisis has taught us the benefit of using innovation, like genetic engineering, to swiftly develop vaccines to treat COVID-19. Farmers and ranchers need biotechnology solutions to solve the challenges they face as well. We need innovations like AquaBounty's genetically engineered salmon allowing us to produce more food with fewer resources. We need genetically engineered pest-resistant or drought-resistant crops. We need new approaches to treat animal viruses like African swine fever. We cannot allow the activists to advocate a return to the slow, conventional route of human vaccine development any more than we can afford to continue to accept resistance and the slow pace of change of acceptance of biotechnology in agriculture. With respect to the evolving shift in consumer acceptance of biotechnology's benefits, we need to capitalize on and extend that mindset to agriculture as quickly as possible.

Biotech crops and animals are a crucial part of the effort to use less land, energy, and water, and fewer agricultural inputs to promote sustainable agriculture, slow climate change, and feed a growing population. Global population is projected to grow to nine billion people by 2050, a 28 percent increase in just thirty years. And yet, even before COVID-19 placed a greater strain on our food supply, 821 million people go to bed hungry every day around the world, and food scarcity is increasing. Lest we think this is a problem for developing countries, thirty-seven million people in the US are food insecure, and eleven million of them are children.

For the agricultural biotech industry, it's time to go on the offensive and proactively shape our future by promoting sound, science-based biotechnology innovations, a progressive and predictable regulatory framework, and transparency and engagement with consumers. These efforts will allow us to address the challenges of human, animal, and environmental well-being globally.

As the company with the first genetically engineered animal approved for human consumption by the US Food and Drug Administration and Health Canada, AquaBounty is committed to continuing to pave the way for others in the pipeline. We take our approval and its significance seriously. The pandemic and its effect on the food supply chain has reinforced the commitment to our mission to be leaders in aquaculture using biotechnology in new ways to deliver game-changing solutions to global problems. As a leader, I identified three important factors to achieving that mission: caring for the team by listening to their concerns and communicating frequently, ensuring strategic clarity and eliminating confusion about what is and isn't critical to our mission and prioritizing ruthlessly because we can't afford to waste resources.

Goethe is often quoted as saying, "Treat people as if they were what they ought to be, and you help them become what they are capable of being." I believe that to be true and see the proof every day in our team members. The pandemic accelerated the development and rollout of our Vision, Mission, and Values. It was a bottom-up effort, as everyone on the team was asked to think through our purpose and the way we want to operate. This commitment to biotechnology and future innovation will help AquaBounty and the industry weather future challenges. Achieving strategic clarity on our purpose and the way we operate is essential. We need every member of the team rowing in one direction. We've created the freedom and expectation to raise issues and concerns if one of them thinks we are off track. Seeing the commitment and talent of our team unleashed is the reason I come to work (virtually) every single day.

We've invested in enhanced biosecurity to protect our people, our fish, and our communities. Many of those enhancements have come directly from the teams on the frontlines. Empowering them to address concerns with appropriate changes has energized the team and improved our practices. That engagement won't stop once the pandemic is behind us, because we've learned that the best ideas often come from those doing the work. We're embarking on a journey of continuous improvement using the practices of Lean and Six Sigma.

Ruthless prioritization requires personal accountability for every member of the team to question whether what they are doing moves us closer to achieving our company mission and meaningfully supports the industry at large. It also depends on the team's knowing they matter by demonstrating we actively listen to one another and provide constant feedback on progress against clear strategic direction. COVID-19 has made us a better organization. Because we are operating more effectively, we can impact the industry and world more effectively.

As Dr. Kenneth M. Quinn, president of the World Food Price Foundation, said, "We have a great responsibility to continue to move science forward and to utilize it in the best ways possible to nourish mankind, especially those suffering every day." That quote motivates me to make sure we have the tools necessary to feed the world and care for our environment. It challenges me to build a team that can know and believe in its purpose and can manage itself effectively. It fuels the passion to know that I am making a difference in the lives of people both in my organization and those who struggle with food insecurity. Let's not let a good crisis go to waste! Instead, let's use it to solve very real problems and help civilization survive into the future.

Sylvia Wulf is the president and CEO of AquaBounty Technologies Inc., leading its growth and commercialization efforts globally. The company produced the first bioengineered animal approved for human food. Wulf is a proven leader driving both growth and improved performance. She serves on the boards of the National Fisheries Institute and BIO.

BUILDING GLOBAL FOOD SYSTEM RESILIENCY THROUGH INNOVATION AND COLLABORATION

Philip Miller

In May 2020, amid the global COVID-19 pandemic, two mothers go to their local market to buy groceries for their families, as they have so many times before. One is an urban dweller in the Northeast United States and the other lives in a small farming community in Kenya.

The American shopper finds that her grocery store is rationing beef, chicken, and milk among customers because of shortages—something she's never seen. Several of the meat processing plants for brands that would have packaged and shipped meat to her grocery store before COVID-19 are closed because of outbreaks among their workers. The consumer dairy supply chain can't keep pace with nationwide panic-buying of milk, among other household staples. At the same time, dairy farmers around her country are dumping an estimated 3.7 million gallons of milk each day[1] because the schools and restaurants the supply was intended for are closed, and the commercial dairy supply chain is unable to pivot production to accommodate consumer-grade products. Infrastructure and labor vulnerabilities in a food supply chain that she had hardly considered have been exposed by the health crisis.

However, the Kenyan shopper knows before she arrives at the market that she could be returning home empty-handed—because she knows her supply chain personally. Since January her neighbors' produce farms and pasture meant for cattle grazing have been ravaged by the largest locust invasion that Kenya has seen in seventy years.[2] By February, 200 billion locusts across East Africa were consuming nearly 400,000 tons of food per day, the equivalent of what eighty-four million people eat in a given day.[3] Because neighboring farmers were already in crisis without access to a pesticide that could control the swarms, the spread of COVID-19 has thinned the local food supply even further as distancing measures and illness brought much production and distribution to a halt. When she arrives, she finds that the limited local supply has led to a price increase for what foods are available—and what she can afford to buy for her family won't last as long as she needs it to.

Innovation access, labor availability, the nature of national policies and regulatory systems, and food movement infrastructure are largely unseen forces that propel our food supply at the global level, and hinder it regionally where shortcomings exist. The COVID-19 pandemic has forced the food and agriculture industries to accelerate conversations and collaborations that will lead to the solutions we need to make our food supply more resilient for all people. And I believe that the role of innovation companies both in convening public and private stakeholders, and leveraging scale to positively affect the sustainability of global agriculture has never been more important.

---

1. https://www.nytimes.com/2020/04/11/business/coronavirus-destroying-food.html

2. https://news.un.org/en/story/2020/01/1055631

3. http://www.fao.org/resilience/resources/resources-detail/en/c/278608

Leading global government relations and regulatory policy for Bayer's crop science division, I work with experts at companies, institutions, and governments on macro issues like these that impact ag technologies. Bayer sells seeds and products that help farmers grow food despite pressure from pests, disease, weeds, and mounting climate volatility. Nearly thirty years ago, I began my career as a biochemistry and molecular biology PhD working on discovering and developing those products in the lab. I would have told you then that the speed of innovation was always going to be key if we were going to be able to produce enough for more people using fewer of our constrained natural resources. What I didn't know then, and have learned from my years spent outside of the lab, is that the speed of innovation isn't usually the problem—instead it's getting innovation into the hands of the people who need it.

I'll give you an example. Fall armyworm (FAW) is an insect pest capable of an incredible amount of damage because it reproduces quickly and can spread long distances rapidly. In production agriculture, it's been a particular problem for maize. More than two decades ago, markets including Brazil and the US adopted *Bt* maize, a crop that's been genetically modified to contain a naturally occurring soil bacterium that controls pests. Outside of the agronomic and economic benefits growers in these countries have seen, preventing FAW from feeding on crops also has a food quality benefit for consumers. These pests can increase the infection of mycotoxins in grain, which can lead to aflatoxins—a known carcinogen for humans. Since 2016, FAW has spread through Africa and Asia, but, because genetically modified crops have not yet been approved by policy makers in most countries in these regions, growers have a limited capacity to slow what is truly preventable damage. Based on 2018 estimates, every year up to 17.7 million tons of maize are lost to FAW in Africa alone. This amount of maize could feed tens of millions of people, and represents an economic loss of up to $4.6 billion.[4]

4. http://www.fao.org/news/story/en/item/1253916/icode/

When policies are not science-based and infrastructure is weak, the people who already had the least food security often suffer most The FAO estimates that by 2050, the world will need to produce approximately 50 percent more food and livestock feed to accommodate population growth and changing diets. We've yet to innovate to match the challenges we face, but I believe we will. The latest advances in data science, predictive analytics, and biotechnology in particular make me excited to imagine what we will accomplish. If precedent holds, though, the larger question will be our ability to deliver that innovation to growers who are hungry for answers, figuratively and literally.

At Bayer, we align ourselves to the global vision of "Health for All, Hunger for None." This ambitious vision is grounded in our core disciplines, health care and agriculture, and is conceivable only based on our global scale and ability to collaborate with bright minds beyond our walls. Some segments of society have developed a mistrust for the private sector, and particularly large multinational companies. But in the past few months I've seen countless companies like ours lining up alongside health care professionals to directly meet immediate COVID-19 treatment needs—and farmers, food chains, NGOs, and governments to deliver innovative solutions to make our food supply more sustainable and reliable around the world. I've seen scale applied for the greater good.

Consider the locusts in East Africa. One of the few active ingredients effective in combating the desert locust is deltamethrin. Bayer has donated 170,000 liters of our pest control product Decis ULV (deltamethrin) to vulnerable countries identified by the FAO, Kenya, and Uganda. By the time this book publishes, impacted communities there will have an added defense against all-consuming locust swarms to better protect local food and financial security as pressure from COVID-19 continues.

Consider also the work we've done to connect farmers more directly to retailers in the food supply chain, particularly in the developing world. What we've seen during COVID-19 is that growers who were better connected and trained in digital resources in countries including Ecuador[5] and India[6] are more easily selling their harvests and maintaining their operations during the disruptions in the supply chain caused by the pandemic.

We're using our worldwide presence to spot potential emerging threats to food security and leveraging our scale to help. We are working with a consortium of experts to solve the Tropical Race 4 (T4) disease crisis in the global banana crop, which threatens the economic and food security of millions in the developing world. We're working with distributors and food retailers to ensure that supply lines remain open and responsive. We're creating strategic stockpiles of seed and crop protection products in key locations for rapid deployment as needed. The COVID-19 crisis has reminded us that our highly interdependent, yet disparate, food system needs to be ready to weather the challenges we can anticipate, and also those we cannot. Creating a more resilient food system requires a concerted, collaborative, and global effort to strengthen our food system against external threats. Improving agricultural sustainability and supply chains will help us withstand and respond to future crises.

Every advancement in agriculture has come about because of a deeper understanding of science and a commitment to delivering innovation. It's my hope that, coming out of a time when science and data were key focal points of our daily existence, we can develop a newfound, healthy trust in science and the entities that pursue it within our society. What if we continued to make decisions based on data and the counsel of experts in their fields? What if we rejected sources of misinformation as threatening to the public good in all areas of our lives, and our policy makers did as well? What if we imagined the experiences of the mother accustomed to a bare market or the grower fighting to keep his small grove of trees alive as we shaped our opinions

---

5. https://www.elcomercio.com/actualidad/alimentos-plataformas-digitales-guayaquil-coronavirus.html

6. https://economictimes.indiatimes.com/news/economy/agriculture/farmers-can-avoid-mandis-by-using-enam/articleshow/74956289.cms

about the technologies that enable better systems of agriculture? What if we maintained the ethos of empathy we've developed for our fellow man in far corners of the world through this historic shared experience, and realized ways we can all contribute to Health for All, Hunger for None? We can, and we should. And you can count on companies like ours to be endeavoring to do so along with you.

Dr. Philip Miller is the senior vice president at Bayer. As chair of the Biotechnology Industry Organization's food and agriculture section, Miller leads a diverse community of stakeholders to develop and advocate for global agriculture policies that will help feed a growing population. He also serves on the Crop Life International strategy council and on advisory committees for the USDA, NIH, and the National Academy of Sciences.

THE PANDEMIC FROM THE FRONT SEAT OF AN AI COMPANY

Alex Zhavoronkov and Evelyne Bischof

To be able to react more efficiently and quickly to future pandemics, to promote global collaboration, and to avoid finger-pointing, it is important to carefully reconstruct, review, and record the timeline of the COVID-19 outbreak from the perspectives of all key stakeholders. On December 26, 2019, I arrived in Shanghai to work on a partnership deal with a pharmaceutical company in China. I used well-populated high-speed trains and subways, and shook hundreds of hands. On January 5, 2020, the WHO published the first news of the coronavirus outbreak.

On January 6, I flew from Shanghai to San Francisco to attend multiple events surrounding the world's largest biotechnology and pharmaceutical convention, the thirty-eighth J.P. Morgan Healthcare Conference, between January 13 and 16. This conference attracted thousands of biopharmaceutical industry executives. In other words, the entire pharma world gathered in one city. Meanwhile, on January 12, China publicly shared the genetic sequence of "the novel coronavirus," later named SARS-CoV-2, and on January 14, the WHO provided a detailed outbreak report. Yet, the debates at J.P. Morgan focused on routine topics, such as the potential for AI in biomedical discovery and development, the importance of digital technologies, the use of AI for progressing research in multiple disease areas, and drug pricing; there was no mention of the coronavirus. It was only when the event was ending that some executives—those planning to go to China—started to wonder whether they would be safe. No one was talking seriously about repurposing their existing drugs to combat COVID-19, or engaging in drug discovery efforts.

But quickly, the epidemic's widespread ramifications became increasingly predictable. On January 23, Wuhan went into official lockdown, just two days before the Chinese New Year, drawing broad criticism from Amnesty International and Western media as a human rights violation, bringing the outbreak into the public spotlight. Around January 24, I proposed to Insilico Medicine board members, several investors, and employees a possible project on the disease—which was not named COVID-19 until February 11. While the board was in favor, some key biotechnology investors and advisers were skeptical. Within another month, the isolated epidemics coalesced into a worldwide pandemic, as classified by the WHO on March 11. Until then, most experts assumed that operations could continue as normal, just as they had during the SARS and Ebola outbreaks. Neither of these previous events had resulted in commercially viable opportunities to develop effective treatments.

Internal efforts at Insilico Medicine

While Insilico Medicine is focused on AI and comprises multiple interdisciplinary groups, we have five main divisions: software development, drug discovery, biology, chemistry, and deep learning theory. After a brief brainstorming session with the leaders of these divisions, we decided to prioritize our projects by the phase of the disease, taking into consideration abundant information from China.

We decided to pursue these strategies:

1. **Generative chemistry**: monthly "sprints" to generate novel molecules for key viral proteins using a generative chemistry pipeline starting from the 3C-like main viral protease (3CLpro).

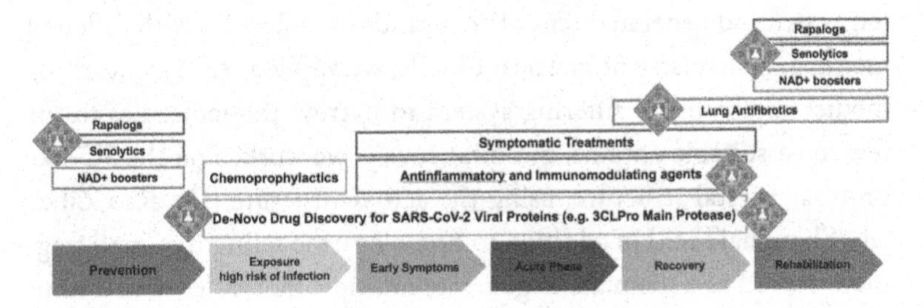

Figure 1. Insilico Medicine COVID-19-related projects by the phase of the disease

2. **Drug repurposing**: repurposing our novel antifibrotic agents to restore lung function and prevent/reverse pulmonary fibrosis. We also proposed a preventative approach using geroprotectors, specifically rapamycin, to boost the immune systems in the elderly.
3. **Repurposing and retraining deep biomarkers of aging**: testing our AI-biomarkers of aging as predictors of susceptibility to infection, severity, and lethality.
4. **Software**: the release of a COVID-19 version, COVIDOMIC, of our Pandomics.com system to be freely available for COVID-19 research, focusing on patient stratification as well as field research. We also decided to expand the software with microbiome analysis tools.

Generative chemistry

On January 26, I got an okay from key members of the Insilico board and other team members, and began work on COVID-19. We quickly identified 3C-like protease as a prime target and applied our generative chemistry platform, which handles multiple scenarios with varying amounts of available input data, to generate 100 molecules targeting the SARS-CoV-2 3C-like protease. We chose a homology modeling approach, using the protein sequence to model several variants of the binding pocket, generating a number of template molecules likely to bind the protein. Next, we expanded the chemical space using these templates and generated tens of thousands of molecules with different medicinal chemistry properties. Finally, we applied our deep learning medicinal chemistry filtering system to narrow the molecules to the few most suitable options. Simultaneously, we worked on the 3C-like protease crystal structure, using the actual structure (Dr. Rao Zihe's lab, ShanghaiTech) in addition to homology modeling. Its small fragment served as a template to generate multiple molecules with a similar structure using ligand-base.

Here, we faced another challenge—all our synthetic chemists were quarantined in Wuhan until the end of March! So instead of performing the synthesis ourselves, we published the 100 synthesized molecules via a dedicated page on our company's website[1] and on the ResearchGate preprint server. This resulted in numerous inquiries from medicinal chemists. After regaining our chemistry capabilities, we decided to synthesize and test a few molecules from these generations.

Repurposing of antifibrotics, senolytics, and rapalogs

One core drug discovery effort at Insilico Medicine focuses on lung and liver fibrosis with novel panfibrotic targets and molecules. We immediately looked at repurposing these for COVID-19 rehabilitation and expanded our preclinical efforts.

We also explored a nontraditional, novel, and more controversial drug-repurposing approach. Instead of looking for known molecules that interact with viral proteins, we looked for molecules that could combat immunosenescence, especially in the elderly and the comorbid population. As initial patient data emerged, it became obvious that the infection rates, severity, and lethality are much higher in these two groups, while younger people experienced milder symptoms and faster recovery. With this background, we postulated that immunomodulators are urgently needed to rejuvenate the immune systems of the elderly, making them more resistant to infections and improving their ability to mount an immune response to possible vaccines. This AI-supported, longevity-based idea has thus far been well-accepted among

1. https://insilico.com/ncov-sprint/

scientific, clinical, and public circles. We hope this idea will extend beyond the COVID-19 era, steering research toward preventative and therapeutic strategies to increase response rates in the elderly toward various diseases, including infectious, autoimmune, and tumor diseases, among others.

Collaborations and governmental support

The multifaceted nature of the COVID-19 pandemic demands global, multilateral collaborations to rapidly advance science and technology. Unfortunately, most governmental funds, pharmaceutical companies, and even special programs in innovation centers lacked the flexibility to refocus and support AI-powered companies specializing in drug discovery for COVID-19, in contrast to other fields. Insilico operates in six countries; however, substantial national support for our COVID-19 initiatives remained absent. Such lack of practical promotion might result in a scarcity of effective drugs targeting SARS-CoV-2, of dedicated AI software systems for multiomics drug discovery, and of comprehensive data for machine-learning exploitation.

Big pharmaceutical companies either started repurposing their drugs for COVID-19, developing vaccines, or donating resources for disaster relief. However, their AI departments produced no tangible papers or products quickly enough; neither did they engage with the key AI companies. This demonstrates the current state of AI in drug discovery and this trend is unlikely to change. It is up to the start-ups to demonstrate the real value of AI.

However, in academic circles the situation is dramatically different. COVID-19 resulted in massive academic collaborations without any funding. Very often these consortia or individual members received substantial computing resources from Amazon, Alibaba, Baidu, Google, Nvidia, and other cloud providers who should be recognized for these efforts. Insilico quickly became part of several such consortia.

In early 2019, Alan Aspuru-Guzik and I discussed the possibility of a viral or bacterial pandemic and the need to create a mitigative, multidisciplinary task force utilizing the latest advances in AI ("Anti-infective AI consortium"). Our initial idea to show a proof of concept in target identification and small molecule generation to inhibit bacterial growth was outpaced by an MIT group co-led by Regina Barzilai. We planned to demonstrate this concept in viruses, but the pandemic prevented us from pursuing this further at the time. Aspuru-Guzik's lab consecutively initiated MyTrace.ca—contact tracing with Bluetooth and GPS via a mobile app. He and many other leaders formed working groups, gathering scientists to work transparently both independently and collaboratively. This model is likely to continue past the pandemic.

AI drug discovery and the clinic

Insilico has consistently worked to engage with physicians. In China, these efforts have proved fruitful, as the country's permanent goals to evolve and invent, coupled with the COVID-19 crisis, represent an unprecedented opportunity to test AI and precision medicine solutions for public health.

Clinicians continue to espouse mistrust of innovative technologies for drug discovery and repurposing. These resentments are mostly rooted in a lack of knowledge about rapidly evolving AI and machine learning. As medics (EB), we fixate on the argument on changeable, adaptive biological systems, and to *a priori* scrutinize tech visionaries. This view fails to recognize that medicine has both a human side, as well as a scientific and technical side—being the science of human nature.

The pandemic disrupted the dismissive, skeptical attitude of clinicians, and facilitated an openness to acknowledging the utility of rigorous scientific AI tools. There is now a growing effort to integrate new technologies into the clinical practice, commencing with fast AI-supported identification of drug targets. Further, there is a new appreciation of AI as a driving mechanism of response to pandemics.

The purely empirical approach has proved suboptimal, with hundreds of studies thus far revealing no therapeutic benefit and with initially promising drugs leading to severe adverse reactions and lethalities.

What will the future hold? Hopefully, a realization that AI and the clinic are inseparable entities, with the patient at the center of attention. Precision medicine cannot be achieved in any other way. Potentially the most significant transformation of all will be the shift toward a paradigm of inclusion of both human and artificial intelligence.

Lessons from China

The COVID-19 pandemic revealed China's capacity to rapidly act and react, create, and innovate. Hopefully, the global community will appreciate the efforts of the scientists, doctors, and the government, rather than criticize or blame China.

The work of Chinese scientists and government laid the foundation and set an example for the world, bringing pivotal repurposing candidates, clinical trials, and a brilliant public health response. China's leaders from all domains were the first to face the novel coronavirus, operating with limited information, including the possibility of asymptomatic transmission. And though their response, including the lockdown of Wuhan and Hubei provinces and efficient contact tracing, was abundantly criticized by human rights advocates and Western media, these techniques proved to be effective and were later implemented in other countries. The scientific community in China was also the first to respond. Within days of the outbreak, hundreds of COVID-19-related WeChat groups formed, some specifically dedicated to relevant scientists. Massive community resources emerged and became available. In January, the Global Health Drug Discovery Institute (GHDDI), an organization founded by Tsinghua University, the Bill and Melinda Gates Foundation, and Beijing municipal government, set up a GHDDI-AI Lab portal on GitHub, gathering all available knowledge, databases, and communities, facilitating cooperation and discussions.

Unfortunately, many scientists have no established contacts with key Chinese labs and are unaware of their research activities. In the post-COVID-19 era, we need closer East-West collaborations, courteous professional relations, and integration of research activities with Chinese scientists, whose data and expertise often surpass the West. Infectious diseases know no borders; so should collaborative efforts.

Conclusion and perspective

The pandemic has caused many social, economic, and political disruptions. Hopefully, it will also disrupt the attitude of ignorance ("this is not my problem and we should not be spending the resources") that existed before COVID-19 and still does, at almost every level. The pandemic highlighted the many inefficiencies in drug discovery and development within the pharmaceutical industry. Both traditional and AI-powered methods failed to provide effective repurposing candidates, novel interventions, or vaccines for several months after the start of the epidemic.

As the pandemic subsides and the dramatic economic impact becomes apparent, it is important to avoid pointing fingers, which can hinder information and resource sharing. Instead, we should focus on fostering closer collaborations between Chinese and Western scientists. COVID-19 and other infectious diseases know no borders and nationalities, and only truly global efforts will lead to effective prevention, drug discovery, and development efforts, and technologies required for rapid responses. Many of these enabling technologies will also help us develop better medicines for noninfectious diseases, which result in a substantial global death toll.

To boost the economy in the near term, avoid the loss of tens of trillions of dollars, and prevent the loss of hundreds of thousands of lives, policy makers and industry leaders must set ambitious goals, like those set by the Apollo program, focusing on increasing the speed of drug discovery and development, and refocusing global efforts on biotechnology.

Alex Zhavoronkov, PhD, is the founder and CEO of Insilico Medicine, a leader in artificial intelligence (AI) for use in drug discovery. Since 2012, he has published over 120 peer-reviewed research papers.

Evelyne Bischof, MD, is an internist, associate professor at the Shanghai University of Medicine and Health Sciences, and physician at the University Hospital Basel, a consultant to Insilico.

LEADERSHIP

LEADING WITH VALUES THROUGH COVID-19 AND BEYOND

Jeffrey M. Solomon

All companies have values. For biotechnology companies, those values are often deeply intertwined with the mission of developing breakthrough therapies, making life science unique: it exists largely to help others, including those with little hope. No surprise, then, that these companies are leading the charge on developing solutions for testing, tracking, and managing COVID-19 and preventing future outbreaks of diseases.

In times l of crisis, leaders of all companies also must take care of their most valuable asset—their people. To keep people safe, a company's ideals, which normally operate in the background like a compass pointing out the organization's direction, should be front and center, acting as a roadmap to navigate uncertainty. In the early days of the outbreak of COVID-19 in the United States, Cowen found itself in the position as a necessary "first mover," requiring us to put our values into action in new and even more tangible ways.

Two days after our annual health care conference in Boston (March 2–4, 2020), where we hosted 2,200 people from across the health care industry, we were informed that two executives from Biogen, who had presented at our conference, were ill and going to be tested for COVID-19. At the time, there was only one confirmed case in New York and no other known public cases in the Boston area. Our senior leadership team met numerous times, virtually, over the following weekend to discuss the health and well-being of our conference attendees, including our employees. We were in close contact with Biogen and learned that, in addition to the two executives who attended our conference, several members of its senior management team had become ill after Biogen's senior leadership conference in Boston the week before. The COVID-19 health threat became very real.

We grasped the implications for our attendees and for the potential spread of the virus within the entire health care investment community. We questioned whether we should have gone forward with the conference, even though we had sought advice from numerous attendees, took extra precautions to limit spread, and had only a few dozen cancellations. We were concerned about the health of our attendees as well as the long-term implications for our brand and our franchise.

We knew this could be a "Tylenol moment" for Cowen: our highest-profile conference, our biggest industry sector, and a contagion risk to 150 of our employees and more than 2,000 attendees. To formulate our plan of action, we contacted several government health officials, but often found them to be lacking tangible advice on how we should proceed. Official guidance was that only those in "close contact" should quarantine, and by our official conference records, that meant only a few people. However, a senior Massachusetts health official had a different take, which ultimately led to us taking bold actions as a "first mover" in the crisis response.

For Cowen's management committee, our decision process boiled down to one thing: what would happen if an outbreak from our conference occurred because we didn't decide to have people work from home and advise attendees to do the same? The answer was clear.

On Sunday, March 8, 2020, several days before the NBA and the NHL postponed their seasons and the NCAA canceled March Madness, we made an unprecedented decision to instruct 150 Cowen employees to work from home and initiated a plan to migrate 80 percent of our workforce to home by the end of the week.

Even though there were only a few confirmed cases in the United States at the time, we saw the storm coming, so we communicated openly with all conference attendees about our decision. We realized that despite any perceived risk to our franchise, we not only had an obligation to inform others, but also to advise and assist them in any way we could with the information we were learning and the actions we were taking. Over the next two weeks, we became an example of what shelter in place looks like across a public company. By the time most government shutdown orders were implemented, we had our own anecdotal evidence about the effects of social distancing that we could share with other companies.

For us, as an independent investment bank and financial services company, the next month became a real-time experiment in remote working and self-isolation during a time of extreme market volatility. By March 13, we had almost 1,000 people globally working remotely. Our actions at the time were well beyond what were then the federal government's and the CDC's recommended precautions. And we did it without missing a beat.

We would not have chosen this first-mover position, but given the circumstances, we embraced the responsibility, including this one: to share how our values became a real-time guide for our decisions and actions.

VEST: The Cowen Values

At Cowen, we view everything through the lens of our values of **vision, empathy, sustainability**, and **tenacious teamwork**—or our acronym, "VEST." From this perspective, there was no doubt as to what we should do.

- **Vision**: In early March, we envisioned a worst-case scenario: the virus spreading throughout our entire organization and the health care industry unless we took decisive, precautionary measures. Everyone's health and safety far outweighed any potential risk of business interruption or concern that we might appear in the public eye to be panicking. Because we followed our vision, we realized a best-case scenario: taking care of our collective health so we could see to the needs of others. Thanks to foresight and planning, we experienced a seamless transition not only to remote working, but also to ensure that the incredible trading volumes (more than four times the daily average, every day during this period) were adequately processed and settled.

- **Empathy**: As the connective tissue of our organization, empathy enables us to function at a high level, even though we were physically separated. In times of stress, empathy also provides a sense of mission and greater purpose—of knowing that we are truly present for each other and our clients. We put empathy into action with our corporate clients who suddenly needed to think through contingency planning and to strategize confidentially with trusted advisers at Cowen about specific scenarios they were facing—particularly around raising capital. We focused our attention on the community, using our network to help others in need. This included raising money for health care workers at Mount Sinai Hospital and working with a company in which our investment management division has an investment to provide much-needed iPads to hospitals.
- **Sustainability**: As a central tenet, at the core of the long-term health of the firm and the importance of building and maintaining relationships, sustainability enhances resilience. This value spurred us to ensure that we were bringing the full weight of the firm—from research to banking to trading to investment management—to every challenge and opportunity. Our business operations team truly shone by executing on our business continuity plan without hesitation. Without them, nothing we do for others would have worked because the firm could not have been sustained.
- **Tenacious Teamwork**: Working together is standard operating procedure. During periods of elevated stress and dislocation, it's mission critical. Tenacious teamwork allowed us to demonstrate to one another and our clients that "we're here for you." In research, Cowen was among the first to provide insights across the

> firm and to our clients on the projected impact of COVID-19, including diagnostics, therapies, and vaccines. We convened global experts to help clients make sense of an evolving pandemic, from the shutdown of the global economy to fiscal and monetary responses. Now, amid discussions of how best to reopen the economy, we see it is as our responsibility to provide expert opinion and market perspective.

Once again, we draw on our values as we look ahead—relying on our vision, acting with empathy, focusing on sustainability, and working together in tenacious teamwork. This allows us to take a values-based approach into the future. Our advice to clients as of this writing is:

Build multiple scenarios into your operating models going forward and stay well financed

Keep in mind that, just as the rolling shutdown occurred, we are likely to have a rolling start-up with those areas hit less hard getting back to work more quickly. For many of us in urban centers like New York, Boston, San Francisco, and London, it probably will take longer than in other cities. More than anything, make sure that your business is well financed in this environment as things could take longer to recover.

Be nimble

For professionals in the biotechnology industry, this means adjusting their plans based on information about how and when clinical trials might continue and the FDA's new processes as they prioritize COVID-19 testing, therapies, and vaccines. Also challenge the current framework for decision-making. If we have learned one thing over the last few months, it is that many assumptions that underpin our strategic planning have become far less certain. Challenge foundational operating principles as you consider new potential, post-virus paradigms.

Do your best to slow things down

Time is compressed, and we are all inundated with data. That's why every day feels like a month, and every week feels like a quarter or maybe a year. But we can keep it simple by filtering out the noise and getting as much clarity as possible.

COVID-19 will eventually peak and decline. The life sciences community will see to that. However, the lessons learned, particularly around the importance of values, should endure into the new normal—and beyond.

Jeffrey M. Solomon is chair and CEO of Cowen Inc. As a passionate advocate for small companies and emerging businesses, he was appointed vice chair and an inaugural member of the SEC's Small Business Capital Formation Advisory Committee. He also serves on the board of the American Securities Association. Previously, he was the cochair of the Equity Capital Formation Task Force and a former member of Committee on Capital Markets Regulation.

Cowen Inc. is a diversified, full service financial services firm and for the last two years has been named IFR's (International Financing Review) US Mid-Market Equity House of the Year. Cowen consistently differentiates its practice by supporting companies through their growth stages and beyond; its biotech expertise, combined with a sustained period of health care issuance, specifically biotech IPOs, continually establishes Cowen as a leader in a sector where its superior health care research coverage is highly valued. With over 100 years as an independent-minded investment bank, Cowen has consistently empowered its clients' success by providing solutions and delivering a depth and breadth of insights fueled by experience and a strong record of success.

FASTER, BETTER, AND CHEAPER? THE PANDEMIC CONUNDRUM

Julie Louise Gerberding

> "If the virus moves faster than our scientific, communications, and control capacities, we could be in for a long, difficult race. In either case, the race is on. The stakes are high. And the outcome cannot be predicted."
>
> —*Julie Louise Gerberding,* New England Journal of Medicine, *May 2003*[1]

In 2003, the SARS coronavirus ultimately lost the first race involving this emerging class of human pathogens, due in large part to the heroic containment efforts in health care settings that quenched transmission. Unfortunately, the current SARS coronavirus, SARS-CoV-2, is proving to be a much more formidable foe. The pandemic infected more than 4.2 million people and contributed to the loss of almost 300,000 lives in less than six months from its first detection—and is far from contained.

The race is on. Today the virus is far ahead of us, but the biopharmaceutical industry is closing in faster than we imagined possible. Already

1. https://www.nejm.org/doi/full/10.1056/NEJMe030067

more than 130 treatment candidates are in various stages of development. Likewise, more than a dozen vaccine candidates are already in the early phases of clinical trials, and some are expected to rapidly progress to the next stages.

Time truly is of the essence. Not only the are the lives of people around the world in jeopardy, but the global economy is also at stake. We must be **faster** than we ever imagined possible. We must speed to develop effective therapies that reduce the mortality rate; speed to find interventions that prevent the development of the more severe stages of disease among those with primary infection; and above all, speed to develop an effective and safe vaccine that prevents infection and transmission during this first pandemic wave and beyond. The collaborations that have emerged to accelerate the timelines for development of the pandemic countermeasures are unprecedented. Fueled by government funds, private sector investments, philanthropy, and regulatory innovations, obstacles are being cast aside to help overcome every conceivable barrier to progress in combating the pandemic.

Speed is necessary, but it alone cannot win the race against this deadly pathogen. We must also create **better** solutions to significantly reduce the morbidity and mortality of infection. More important, efficacy must be achieved with an acceptable safety margin, especially among the diverse populations of vulnerable people who need treatment. Predicting which of the many candidates will provide better efficacy and safety is not easy in the early stages of their development and has motivated the emergence of innovative private-public partnerships for scientific collaboration. Projects like the National Institutes of Health's "Accelerating COVID-19 Therapeutic Interventions and Vaccines" (ACTIV) partnership aim to build a "wise crowd" of experts, including those in the biopharmaceutical industry, who work together to establish the profile of desired medicines and vaccines, develop harmonized protocols for studying their value, and then make early choices about the most promising candidates that merit at-risk investments in accelerated development and manufacturing capacity.

Science is on our side as we approach these challenges, but these are still early days in the pandemic and there is much to be learned.

Despite the plethora of clinical experience and rapid dissemination of new observations, we are also in the early stages of understanding this pathogen. Its mechanisms of infection, transmission parameters, interactions with the host immune system, multiple clinical effects, and long-term consequences are areas where more epidemiologic, clinical, and laboratory investigations are needed. Already our understanding of COVID-19 disease has evolved from the early recognition of severe lower respiratory disease, to the more common diagnosis of mild upper airway disease. Broader systemic effects are now evident as well, including involvement of the blood vessels and blood coagulation abnormalities which may underlie many complications, including embolism, stroke, heart attacks, organ failure, and even "COVID toes"—a painful condition probably caused by inflammation of small blood vessels. Likewise, the recent association of pediatric SARS-CoV-2 infection with a severe immune disorder that appears very similar to Kawasaki disease, another devasting disease association with blood vessel inflammation, is an ominous sign that children may be at higher risk for disease than we initially thought. As the epidemic progresses and more patients are studied closely, it is likely that other less common manifestations of infection will become evident. Hopefully the ongoing studies will broaden our understanding of how this virus causes disease and help identify potential new therapeutic targets that could be better than conventional antiviral therapies alone.

Finding a faster and better set of countermeasures to fight a new infection is challenging enough, but when dealing with a global pandemic of this magnitude it is also imperative that solutions are **cheaper**. We must ensure that effective medicines and vaccines are affordable, safe, and can be made available on a global scale. Never in the history of humankind have we been tasked with finding an affordable vaccine for everyone. To put this challenge into context, consider that today we are not even fully successful in immunizing the world's birth cohort for vaccine-preventable childhood diseases, despite decades of advocacy, effort, and investment. Or that despite our long-term awareness of the threat of an influenza pandemic, the annual global supply of influenza vaccine is far less than two billion doses, and that most people in resource-limited countries have no access at all.

The cost of manufacturing a dose of pandemic vaccine or drug is expected to benefit from the economies of scale that spread the fixed costs of development and manufacturing over a large volume of doses. But affordability also depends on the resources needed to distribute medicines and vaccines to local communities and then safely administer them to the people who need them. Access requires a health system with the capability to manage innovative medicines that may require refrigeration, injection or infusion, or development of products that bypass these requirements and are more suitable for mass distribution in community settings. Uptake requires a modern health system that works, a trained health care workforce willing and able to administer innovative treatments, and a populace with sufficient trust in their value to step forward and receive them. These are tall orders in most countries, and very difficult challenges in resource-limited environments among people with little or no experience with novel medicines.

Can we end this pandemic? Yes—but we must defy an old adage and simultaneously be faster, better, and cheaper. Achieving two out of these three criteria is not good enough. We must be much faster. We must provide significantly better medicines and vaccines than those than those previously imagined. And, above all, we must assure that our solutions are cheap enough to be equitably available to the global community. This is a daunting but doable mission. Indeed, it embodies the core values of our biopharmaceutical industry and the purpose that motivates each of us to commit our lives to our profession. This is our finest moment, and together, I have no doubt we are up to the challenge.

Dr. Julie Louise Gerberding is executive vice president and chief patient officer at Merck & Co. Inc., where she is responsible for global policy, strategic communications, population health, and other functions. She joined Merck in 2010 as president of vaccines and was instrumental in increasing access to the company's vaccines among people around the world.

Previously, Gerberding was director of the US Centers for Disease Control and Prevention, where she led the agency through SARS and more than forty other emergency responses to public health crises. She serves on the boards of Cerner Corp., BIO, Case Western Reserve University, and the MSD Wellcome Trust Hilleman Laboratories, and she cochairs the CSIS Commission of Global Health Security.

A CHANCE FOR THE PHARMACEUTICAL INDUSTRY TO RECONNECT?

Jeff Berkowitz

When I joined a large retail pharmacy chain after well over a decade as an executive at a large pharmaceutical manufacturer, I thought I understood the pharmaceutical supply chain—how a drug is developed, approved, priced, distributed, delivered to a patient, and paid for.

I was dead wrong. Within two weeks of arriving, I realized I knew very little about how retail pharmacy worked, how it made money, or how drugs flowed through the system—even though my new employer filled one in every four prescriptions in the US, including the very drugs I shepherded to commercialization. Similar story when I transitioned from retail pharmacy to join the leadership team of the largest payer/pharmacy benefit manager (PBMs): despite being responsible for payer relationships throughout my career, I discovered again that I knew little about the intricacies of their business and their challenges. The "partnering" with other health system stakeholders that my pharmaceutical colleagues and I would talk about was, in many ways, just that: talk. The only intersection was our goal of getting our drugs through the channels to patients.

Meanwhile, these other health system stakeholders have been really partnering with each other—by massively consolidating. Three groups, United/Optum, CVSHealth (Aetna), and Express/Cigna, now affect the delivery and payment of health care for more than 75 percent of the US population. They combine the largest insurers with the largest PBMs; they own the three largest specialty pharmacies; have aligned formally or informally with large retailers; created strategic relationships with the three large drug distributors; and increasingly own health care providers. The implications for health care overall are huge, but from pharma's distant window it is this: these groups have enormous negotiating power on drug pricing.

Manufacturers' gross-to-net reductions—the difference between the list price a manufacturer sets for a drug and what payers actually pay—has more than doubled since 2012. The average discount for branded pharmaceuticals runs at nearly 50 percent.[1] Payers wield many weapons to extract these discounts, including the ultimate one: excluding a drug from coverage. CVS and Express Scripts, two of the largest PBMs, now exclude 282 and 296 drugs, respectively—up from 154 and eighty-five just a few years ago.

In response, pharma has focused on those drugs whose coverage payers and PBMs can't easily restrict: specialty drugs for serious diseases of smaller populations, and particularly oncology and rare disease therapies. Specialty drug spending has now surpassed spending on drugs for major chronic diseases, moving from 26 percent of drug spend in 2009 to nearly 50 percent in 2019.[2] Oncology and rare disease drugs dominate the drug development pipeline. In four of the past five years, over 40 percent of first-product launches were orphan drugs; the late-stage oncology pipeline included 849 molecules in 2018, up 77 percent since 2008 due to the increasing number of targeted therapies.[3] The increase in rare and specialty drug discovery and development has come at the cost of the development of new primary care

1. Pharma ChartPack.
2. Express Scripts Drug Trend Report.
3. IQVIA Global Oncology Trends 2019.

drugs, on which payers can exercise their greatest negotiating leverage. But for the other health care stakeholders, in many ways, primary care is also our society's area of greatest need: cardiovascular disease, diabetes, and chronic respiratory conditions will, over time, kill far more Americans than will COVID-19. During my time in the world of retail pharmacy, PBM, and payer, the discussion was always focused on those major disease states and how to manage them across broad populations rather than potential cures for the top rare diseases.

Not that pharma's success in creating therapies for less prevalent diseases hasn't created enormous medical benefits. Drug companies have made great progress in treating late-stage cancers; we can now save children who, just a few years ago, would have been doomed by conditions like cystic fibrosis.

But these advances haven't been generally lauded. Reputationally, the drug industry has never been seen so negatively[4]. The public at large are subject to near-constant rhetoric about high drug prices and high-priced lobbyists—but, one result of pharma's shift from large-population medicines is that only a few consumers see the benefit of much of the industry's R&D output today.

The COVID-19 pandemic has given pharma a chance to reconnect with the American public. The industry's importance and its capabilities have never been clearer. Drug companies have, in just a few months, created and started testing more than 300 COVID-19 vaccine and drug candidates.[5] To get all this done, they are collaborating with one another in unprecedented ways—putting competition and differences aside and partnering on clinical trials and sharing data.

Payers—whose reputations are barely better than pharma's—are attempting to do their part too. All the top insurers have expanded access to telehealth services and cut patient costs for them; eliminated patient cost-sharing for COVID-19 related diagnosis and treatment; and waived or at least increased refill limits on prescriptions. They've

4. https://news.gallup.com/poll/266060/big-pharma-sinks-bottom-industry-rankings.aspx.

5. Biocentury: "Daily Chart: COVID-19 Programs Underway."

done this because they know patients will avoid testing and treatment if their costs are too high (and consumer out-of-pocket spending has been skyrocketing as payers put more of the cost burden on the patients' shoulders). Payers know that by removing barriers to care, outcomes will be better.

Both pharma and payers will undoubtedly get a reputational boost from their activities. But perhaps the lessons learned during the pandemic can lead to pharma's reconnection with the broader health care ecosystem in a more profound way—inventing new treatments for widespread disease is not merely medically but societally important. Likewise for payers, realizing that lowering patient out-of-pocket costs—removing barriers to care—will actually improve patients' access to medicines, and benefit society as a whole. I would submit that one way to lift all boats in this rising tide would be to reconnect the disparate efforts for even greater impact.

For example, pharma has developed a compelling but complex set of patient support programs, mostly focused on copay assistance, to provide more robust access to drugs for patients who find it increasingly difficult to afford them even when covered by payers. At the same time, payers continue to shift more and more of the cost burden of those drugs to the same patients through high-cost deductible insurance programs and ever higher copays. The two sets of action are, managerially, completely disconnected—at least in part because pharma has itself been disconnected from the vastly more consolidated payer world and in part because payers use copays to steer patients away from one brand to another—or away from branded therapy entirely. And, as stated previously, doing so has simply increased the pressure on pharma to move away from primary care where payers have the most leverage but where the total pharmaceutical benefit, and reputational impact, could be greatest.

I don't pretend this challenge is easy to solve. I suspect that government incentives may be part of the answer (after all, they've been very helpful in encouraging pharma's focus on oncology and rare diseases). To get this process started, pharma can build on the foundation of cooperation created during this pandemic to reconnect with the broader, consolidating ecosystem that has largely left it isolated with a determination to solve problems collectively. Machiavelli suggested that "one should never waste the opportunity offered by a good crisis." Let's use the COVID-19 crisis to plug back in.

Jeff Berkowitz is the CEO of Real Endpoints, a leader in providing insights, support, and tools to strengthen access to pharmaceutical innovation for an evolving health care landscape focusing on value-based arrangements, patient services, and access support for manufacturers and payers.

His perspective has been shaped by a career spanning most key verticals in health care with executive roles at UnitedHealth Group, Walgreens Boots Alliance, and Merck. Berkowitz, a lawyer by training, is a recognized expert in market access and reimbursement for pharmaceuticals and serves on the boards of several publicly traded pharmaceutical companies including Lundbeck A/S, Esperion Therapeutics, Zealand Pharmaceuticals, and Infinity Pharmaceuticals.

WILL COVID-19 TRANSFORM THE POLITICAL LANDSCAPE?

James Greenwood

If you were watching Chinese television at the stroke of midnight on New Year's Eve, 2019, you would have witnessed Shanghai's spectacular multicolored, illuminated, computer-synchronized drone show—a stunning technological display of nonpolluting high-tech "fireworks."

But the Chinese government was not being honest. The event had been filmed three days earlier to be sure that any glitches could be hidden from the public, so as not to embarrass the regime. When citizens of Shanghai stepped outside to see firsthand what was on their television screens, they looked up at an empty sky.

Earlier that day, the Chinese government had announced that doctors in Wuhan were treating dozens of its citizens afflicted with a mysterious pneumonia. As with the drone show, the official news agency failed to mention that the real events had occurred earlier in the month and had been kept secret to avoid public panic.

In the United States, revelers welcomed the arrival of 2020 with the traditional Times Square party, televised performances by Jennifer Lopez and Sting, and fireworks. Far from their minds were thoughts of the bitter congressional and presidential election campaigns that would soon ensue.

Those of us in the biopharmaceutical sector knew well that those campaigns would feature yet another cast of politicians armed with cheap-shot, poll-tested, rhetorical one-liners condemning drug companies and their "skyrocketing" prices. This, even though the overall rate of increase in drug spending in 2019 was just 2.3 percent—in line with the Consumer Price Index. Notwithstanding, among the proposals to be bandied about would be the imposition of International Price Indexing and inflation caps on medicines.

This political grandstanding is profoundly disheartening to us. The unprecedented level of political condemnation and the resulting poorly conceived stack of federal and state legislative proposals aimed to attack the industry come just when our scientists are developing the most game-changing medical technologies in history. New gene, cell, and immunotherapies as well as CRISPR-Cas9 gene-editing technology are revolutionizing drug development and offering new hope to millions of patients with cancer, diabetes, Alzheimer's, ALS, Parkinson's, and thousands of rare diseases.

Three weeks into the new year, a thirty-year-old man who had traveled from Wuhan to Seattle was diagnosed as having the new disease, a novel coronavirus disease later named COVID-19. Ten days later, the Trump administration suspended entry into the United States by any foreign nationals who had traveled to China in the previous fourteen days, excluding the immediate family members of American citizens or permanent residents.

On February 18, two weeks before the first COVID-19 death in the US, Dr. Jeremy Levin, chair of the Biotechnology Innovation Organization (BIO), sent an email to our nearly 1,000 member companies. He expressed the goals of our newly formed BIO Coronavirus Collaboration Initiative aimed at "bringing together the brightest minds in our member organizations, along with government public health agencies, thought leaders, and other experts to exponentially amplify what we are currently doing individually." Our goals were to help those who wanted to get involved meet potential collaboration partners and find their way through government agencies and programs to develop solutions for patients as quickly as possible. The expectation was that we could work together with "philanthropic purpose" to benefit patients. Levin wrote, "We have the expertise and talent to be part of the front line of defense against this grave threat, and we are doing so with good intentions for the benefit of humankind." He recruited to chair the project George Scangos, CEO of start-up immunology company Vir Biotechnology. Scangos had previously served for 14 years as CEO of the major biotech company Biogen.

Nearly fifty companies responded that they possessed the scientific foundation, development expertise, and/or the manufacturing capabilities to contribute to the cause. Overwhelmingly, these companies made it clear that their commercial interests would be secondary to their commitment to get medicine to patients as swiftly as possible.

On March 24 and 25, we convened a virtual summit in which more than 500 attendees from biopharmaceutical companies, academia, nongovernmental organizations, and federal agencies, including the Food and Drug Administration, the Centers for Disease Control and Prevention, the Biomedical Advanced Research and Development Authority (BARDA), the National Institutes of Health, and Ambassador Deborah Birx, the White House Coronavirus Response Coordinator.

From the summit came the formation of working groups focused on developing diagnostics, therapeutics, and vaccines to fight COVID-19. Scores of collaborations followed. BIO also set up a Coronavirus Hub to connect companies with excess capacity and resources with those that needed them. This new hub would enable users to post requests for urgently needed items and announce the availability of supplies and capacity. The portal connects in real time through customized and searchable postings.

Additionally, BIO contracted with consultants to learn the extent to which needed capacities and supplies exist globally so they could be shared in a centralized and freely available information repository and to prevent gaps and bottlenecks.

On March 27, President Trump signed into law the Coronavirus Aid, Relief, and Economic Security (CARES) Act, which provided more than $2 trillion to support workers, businesses, states, and localities hit by the unprecedented financial impacts of the suddenly essential closure of the economy. The law provided $3.5 billion to BARDA to provide biopharmaceutical companies with funds for the development and manufacturing of vaccines, diagnostics, and treatments, and for the purchase of these products. We volunteered to help expedite our member companies' access to these funds.

Even as this emergency legislation moved through Congress, we had to defend against amendments that would have required biopharmaceutical companies to charge "reasonable" prices for COVID-19 therapeutics and vaccines. We had no doubt that companies would, without question, make certain that their therapeutics and vaccines would be affordable and accessible to all, but did not want to relegate the determination as to what was "reasonable" to political manipulation.

We tried to include in the CARES package amendments that would 1) have allowed certain patients at the discretion of their physician to self-administer their medicines at home if they couldn't safely get to infusion centers, so they wouldn't have to enter hospitals overwhelmed with COVID-19 patients; 2) make sure that small biotech companies would be eligible for funds under the Paycheck Protection Program; and 3) incentivize innovation of new antibiotics to respond to the problem of antimicrobial resistance. Unfortunately, the negative political environment surrounding the drug industry made it impossible to have these patient-centric provisions included.

The federal government was not prepared for this devastating pandemic. It should have been. Since 2014, I have served on the Bipartisan Commission of Biodefense. The commission is cochaired by Tom Ridge, a former governor of Pennsylvania and former US Secretary of Homeland Security, and Joe Lieberman, a former senator from Connecticut. My fellow commissioners include former Senate Majority Leader Tom Daschle and Ken Wainstein, former Homeland Security Adviser to President George W. Bush. Donna Shalala, former Secretary of Health and Human Services (HHS), was an original commissioner before she was elected to Congress in 2018. She was replaced by Lisa Monaco, former Homeland Security and Counterterrorism Adviser to President Barack Obama.

We convened extensive public meetings in which we heard from the nation's best experts on a range of related subjects. In October 2015, our commission published *A National Blueprint for Biodefense.* We recommended that the vice president of the United States lead federal planning to prepare for inevitable global pandemics and bioterrorist events and develop and implement a comprehensive national biodefense strategy. We called for a system of stratified hospitals with increasing levels of capability to treat patients affected by highly pathogenic infectious diseases. We said that Congress should tell HHS to provide financial incentives for these hospitals to prepare for pandemics and to allow for the forward deployment to major metropolitan areas of medicines, personal protection equipment (PPE), medical devices, and other essential medical supplies from the Strategic National Stockpile. We advised the government to revolutionize development of medical countermeasures for emerging infectious diseases with pandemic potential, to incentivize development of rapid point-of-care diagnostics, and to lead the way toward establishing a functional and agile global public health response apparatus. Had our recommendations been more comprehensively adopted, the impact of the COVID-19 pandemic could have been significantly less costly in lives and treasure.

Now, the entire world population is at risk from the deadly disease. Billions are living under some form of stay-at-home order. The global economy is expected to suffer the worst recession since The Great Depression. Doctors, nurses, first responders, public health professionals, grocery store clerks, and other essential workers toil at the perilous front lines.

There is only one way out of this: biotechnology innovation. Only the men and women who labor in their laboratories, the entrepreneurs who lead the biopharmaceutical companies—large and small—and those who invest in their highly risky enterprises can protect us from the virus and get us back to work, back to school, and back together.

We will succeed. We will develop new diagnostic tools, so we know who among us has been infected and who has not. We will develop antivirals and other therapeutics to treat those with the disease. We will innovate new vaccines that will arm the entire global population with antibodies to kill the coronavirus as soon as it enters our bodies and before it can wreak its damage.

Most of the projects our companies undertake to fight the virus will fail. The companies' and their investors' money will be lost. This is how it always is in the world of biotechnology. But the war will be won. Those companies who still stand will turn to their next challenge.

As we achieve these stunning successes, patients, health care providers, the media, and the policy makers will marvel at each advancement. They will celebrate the lives saved and the economies recovered. And they will rejoice at our longed-for desire to embrace our loved ones and friends, return to work, and join our neighbors at parties, sporting events, concerts, and movie theaters.

But will they celebrate the heroic men and women of biotechnology who will have made this possible? Or will they continue to malign and vilify the world's greatest source of medical progress—America's innovative biopharmaceutical industry? If they persist in disparaging the medicine makers, they will have squandered the opportunity to learn a most valuable lesson. Our ability to prepare for the next—inevitable—global pandemic will be diminished, as will the hope for patients with every disease.

But if, instead, Americans of all walks of life having seen how we labored, how we collaborated and how we applied our genius and imagination to bring the world back from the brink of despair and collapse, they may gain a new and more open-minded understanding of our vision, our mission, and our dedication. If that happens, the miracles will continue to multiply.

The Honorable **James Greenwood** has served since 2005 as the president and CEO of the Biotechnology Innovation Organization (BIO). Previously, he represented Bucks County, Pennsylvania, for six terms in the United States House of Representatives, following twelve years as a state representative and Senator. Before serving in elective office, Greenwood was a caseworker with abused, neglected, and disabled children.

The Biotechnology Innovation Organization (BIO) is the world's largest trade organization representing biotechnology companies, academic institutions, state biotechnology centers, and related organizations across the United States and in more than thirty other nations. Its mission is to advance biotechnology innovation by promoting sound public policy and fostering collaboration, both locally and globally.

INVESTOR PERSPECTIVES

A POWERFUL YET VULNERABLE INTERCONNECTED, INTERDEPENDENT WHOLE

Nina Kjellson

The job of an early-stage investor is *connecting*—people, ideas, patterns—and *spotting*—trends, themes, hurdles—in the name of accelerating innovation. In my case, it is for the sake of speeding the delivery of new medicines or drug discovery platforms. Over my twenty years as a life science venture capitalist, major trends have unfolded: a transformation of the scientific toolkit (new biologics, CRISPR, next-gen crystallography, computational power, etc.); an explosion of the life sciences workforce; and externalization of research and development to contract firms with deep expertise.

Today we partner with founders to build start-ups both as fully integrated companies and as chimeras of internal and external resources. The latter gives us flexibility and cost savings, and access to better expertise than we might be able to recruit to an early venture. However, given how intimate and iterative discovery science is, these contracted resources become deeply integrated team members in places like China, Italy, Switzerland, and the UK.

So an early spot for me in the SARS-CoV-2 pandemic was the impact on our colleagues at Chinese contract research organizations, who became travel-stranded celebrating Chinese New Year. Even before tallying that half or more of Canaan's biopharmaceutical investments have international dependencies, we saw the practical and emotional impact on our teammates facing lockdowns and personal exposure. A reporter called about the effect on our portfolio R&D activities. The effect is real, but my first reflex was the human side.

Then came the first US cases and social distancing recommendations. VC firms are partnerships, not companies. But some firms—like ours—are more family-like than others. Though Canaan operates across four physical offices, two industry segments, and three to four generations, our firm is a strong community. We saw the "distancing" signals in New York, Connecticut, and the Bay Area quickly and—with gratitude for our privilege and flexibility—closed our offices and shifted to work from home even before local ordinances mandated shelter in place. We already run our firm with videoconferencing and heavy e-communication (I even wrote a piece about it early in the pandemic),[1] so we pivoted quickly to focus on how we could support the portfolio companies to do the same and to brace for what's to come.

Twenty-plus years into my venture career, there are certainly prior downturns to reflect on—the whiplash of '99–'01, the heartbreak of 9/11, the 2008 financial crisis. As investors or operators, many of us have been here before with veins full of adrenaline and cortisol, looking at cash runways, organization charts, and business plans to sketch out Plans A, B, and C. These downturn blows almost always come after a period of sustained growth and expansion, meaning the plans being gutting are particularly hopeful ones. The sails are *really full*. It is wrenching. The advice we can give as seasoned investors includes:

---

1. https://timmermanreport.com/2020/03/dos-and-donts-of-staying-connected-in-the-time-of-physical-distancing.

- Contingency planning is power.
- Extending the runway sooner is better.
- There is grace in communicating honestly
- Being generous where you can is a virtue, but you have a responsibility to the business and to all shareholders.

It is heartening to see the leadership, grit, and humanity across our companies and the poise with which executives are stewarding not only their businesses but their culture. This seems especially true in the life sciences, where **humanism is almost as prevalent as genius**.

As VCs, we seek to rally that spirit and remind leaders to take care of themselves to sustain their resilience. We marshal resources. We've created CEO and founder communities for sharing ideas and best practices; held ask-me-anythings; participated in virtual panels; circulated how-tos on layoffs, crisis communications, virtual fundraising, hiring without meeting, SBA loan applications, and nurturing culture in a time of distancing. All the while, we're doing internal impact analyses on the Canaan funds, our own individual portfolios (they comprise our personal track records or report cards, after all), and forecasting changes in potential outcomes.

Our biopharma companies are now planning a return to work in offices and laboratories—beyond the limited "essential" work that they have been continuing to do as permitted under local ordinances. Most discovery-stage companies would say "wet lab" work is at 40–50 percent normal capacity—constrained either by the pace of materials delivery or the pace of conduct of experiments with researchers working in shifts and spaced six to ten feet apart. There is desperation to get back to 100 percent. Estimates are 60–75 percent with wider opening up. Clinical-stage companies have faced a range of realities: those whose protocols require access to a hospital setting, or assurance of an ICU bed in case of a safety event, have pretty much stalled. Those with trials that require few office visits and less complex administration have been less affected and are more quickly coming back online now. Some have accelerated into new COVID-19 treatments. And since a great deal of our portfolio focuses on oncology or other high unmet needs, overall, there is high motivation by all to see patients receive potentially life-saving, last-resort treatments. There has been much collaboration and creativity.

Partnering transactions have mostly continued. The interplay of start-ups and big pharma to drive innovation continues to be a strong industry trend. We've signed two massive collaborations and have two more in the works. The *strategics*—as we tend to call the big pharma companies—have so far resisted taking egregious advantage of their deep-pocketed, COVID-19 upper hand.

In terms of making new investments, we are fortunate to be a well-capitalized, thirty-plus-year-old firm with strong networks and a consistent and disciplined thesis. In our biopharma practice, we are actively looking for new opportunities with strong science and exceptional founders. We think that if the vision is a highly impactful medicine for patients, a good business case will follow. In light of COVID-19 and the public health and global vulnerabilities this pandemic has revealed, perhaps our apertures will be a little bit wider than before to diagnostics, vaccines, and the data and computational infrastructure for detection, coordination, discovery, and development.

Canaan is already a leader in diversity and inclusion, as half our investors are women and half are immigrants or first generation. As a result, we invest in female and minority founders at twice the rate of our peers. But the racial and economic injustice of COVID-19 is a life and death issue and underscores the urgency of our efforts. Whether higher rates of infection among essential workers, those unable to isolate, the homeless, or the incarcerated or the much higher COVID-19 mortality among those with chronic conditions aided and abetted by unequal access to care and negative social determinants of health—this virus is deadlier if you are black, brown, or poor. One friend lost his father in a federal facility with the highest per capita infection rates in the US. Another spent a family Zoom call—with some sixty participants—explaining advanced directives, while in her day job she lobbies that organ function not be weighted as a criterion for ventilator assignment, because that puts an African American man points behind, before the counting even begins. While I have been a leader in our drumbeat for diversity and inclusion, it is not enough. We must invest in—and represent—underrepresented and use innovation to help dismantle institutional barriers to equity of health and well-being. If we don't tear them down, we risk propping them up.

While not the direct domain of a seed and early-stage firm such as ours, it does not escape me that literally billions of dollars are being invested to scale COVID-19 therapeutic and vaccine development and manufacturing efforts. The Milken Institute counts some 100 vaccines and nearly 200 treatments in the pipeline.[2] Public and private entities are funding capacity and procurement at risk, even before evidence of efficacy and safety from various early programs. To this, I say: bet on biopharma! It is the richest armamentarium of technologies and talent in human history. Yes, there will be failures. Yes, there will be shortcuts taken with both gains and harms. There will be profiteering and turf battles. But if even one innovation proves out and we stand ready to manufacture and distribute, the returns in terms of human lives and economic productivity will eclipse the dollars risked.

Personally, I am reflecting deeply on what will be different and what will remain the same on the other side of the SARS-CoV-2/COVID-19 pandemic. The past eight weeks have transformed how I work, how I "meet" with known and new colleagues—each conversation is infused with a bit more authenticity and gratitude for our respective blessings and the good fortune of interesting and meaningful work. Like many others, I've toddled into new relationship and family territory—with much more, less, and different time together. I've loved rediscovering household activities—cleaning, laundry, meal-planning, gardening—previously outsourced or abdicated to long days and a commute. I am sleeping more and exercising more. I live near nature (woods and water) and have been able to access it most days. As I think about our "re-opening," I am thinking about the impact of my old routines and travel on not just myself and my family, but on my broader community and on the environment. How can I preserve some of the beautiful balance

---

2. https://timmermanreport.com/2020/03/dos-and-donts-of-staying-connected-in-the-time-of-physical-distancing.

that I've been given in this strange, unsettling, and existential time? I have certainly found immense focus and productivity working from home, but I also miss my biotech and venture capital tribes—I am excited to see them and to play in this wonderful intellectual sandbox together, *in person. At least some of the time.*

A final common role of a connector venture capitalist is to be our own hub and spoke, so other corona activities for me have included nurturing several communities—our internal team at Canaan; an alumni group of biotechies who climbed Kilimanjaro to fight cancer; a group of interested philanthropists sourcing PPE; my Aspen Health Innovators Fellowship community of extraordinary clinicians and leaders, which has spawned working groups on testing and equity and privacy considerations in the pandemic; a fundraiser to support local frontline workers; career counseling graduates coming into an uncertain labor market; and stewarding WoVen, our advocacy platform for women investors and entrepreneurs. If there is anything this virus is teaching us, it is that we are a powerful yet vulnerable interconnected, interdependent whole.

Nina Kjellson is a general partner at Canaan, an early-stage investment firm. She leads Canaan's Women of Venture platform for female entrepreneurs and serves on the boards of Essential Access Health, Girl Effect, and Oliver Wyman's Innovation Center. She advises Springboard, Nina Capital (no relation), and the Gates Foundation, and is an Aspen Health Innovators Fellow.

BIOTECH'S RESILIENCE: AN EARLY-STAGE VC'S PERSPECTIVE

Bruce Booth

It's difficult to understate the impact of the COVID pandemic on society. Beyond the incredible suffering of patients and the profound loss of life, this crisis is ravaging the global economy in ways not seen since the Great Depression. US GDP in the second quarter is expected to drop by more than 30 percent, as only essential businesses operate in the most expansive voluntary lockdown of the economy in history. The dramatic collapse of the oil markets, and the financial ripple effects of that into other sectors, compound the already serious challenges we face.

As a surprise to some in the face of this dire pandemic, the biotech ecosystem has been remarkably resilient—as has my conviction as an investor. My colleagues and I at Atlas Venture continue to actively work with entrepreneurs—via videoconferencing—to help create the next generation of exciting biotech start-ups, even amid the COVID lockdown.

Despite the dislocations of this crisis, I haven't changed my "first principle"[1] investment thesis for biotech*: if we can positively impact the*

1. https://lifescivc.com/2019/12/venturing-a-perspective-on-the-drug-pricing-debate.

lives of patients by discovering and developing innovative new medicines, the system will reward that risk-taking with superlative investment returns.

With ample time to reflect on what I love about biotech during #socialdistancing and working from home, I thought I'd share three reasons for my bullish optimism in the face of this COVID pandemic: First, biotech is a unique part of the venture capital and business ecosystem, and demand for our medical innovations is inelastic with many traditional economic forces; second, as an early-stage venture investor, I'm a big believer in the strategic nimbleness of start-ups in the face of challenges, as evolving adaptability is in the DNA of many young biotechs; and third, I have complete conviction that innovative and high quality science, and the scientific method, will lead us toward great opportunities for patient and societal impact—even more true now as the only path out of this crisis. Let's tackle each of those in turn.

Biotech beats to a different drummer

Biotech venture investing is the definition of patient long-term capital, and the COVID crisis has only made that more apparent. Even with the unprecedented economic changes, incredible equity and bond market volatility, and the general "risk-off" sentiment, venture investing in innovative biopharma companies appears to be continuing with conviction, helping them advance their therapeutic pipelines. Even the public equity market appears receptive to helping finance emerging biotech companies, with over ten biotech IPOs in 2020 through just the first four months, aimed at advancing high risk but impactful medicines across a wide range of diseases.

This remarkable resilience in the face of this pandemic's dislocations is a surprise to some, but biotech has always beat to its own drummer relative to other venture capital sectors. In early-stage biotech, experimental data the ultimate currency of progress and value, rather than "normal" business metrics (e.g. revenues, EBITDA, etc.). While typical industries are often affected by acute changes in consumer demand (and spikes in unemployment), this isn't really the case with early-stage biopharm, where the principal driver of value is whether or not data generated demonstrates real value to patients. While many other business metrics are elastic (changing) in the face of economic forces, demand for high impact medicines (as measured by data) is largely inelastic—it exists almost regardless of the acute economic backdrop. Loss-making R&D-stage biopharma is where many of these new drugs are discovered and developed today: companies that get funded, and valued, based on data that accrue over years, not weeks and months. Further, even in challenging economic cycles, patients need their medicines. That disconnect from conventional economic cycles is one reason biopharma tends to outperform other sectors during financial recessions.

Start-up DNA encodes adaptability in the face of a changing environment

We like to plan in the venture creation business—envision what a new start-up could deliver and then lay out the plan to get there. But paraphrasing Eisenhower, plans are worthless, but planning is everything. This is true for most strategic planning processes for any organization, but in particular for start-ups in the midst of macro shocks like a pandemic. Start-ups have the benefit of being nimble, free from the burdens of large fixed infrastructure, which enables a broad possible range of responses to external shocks—in short, start-ups can pivot faster to new opportunities, and abandon older concepts or recently disproven ones. Truth-seeking is a hallmark of the best biotech decision-making process.

While biotech as a sector may have differences from other business areas, we are not Pollyanna about the impact of COVID on the sector. Even though financings are continuing and the floor hasn't fallen out of the market, there are ample reasons to be concerned, most of which involve R&D execution. This is where adaptability and keen decision-making become important.

The biggest impacts of the COVID-19 crisis are on R&D timelines and delivering the next set of R&D milestones; in short, it raises challenges to a biotech's ability to get key value-creating experimental data in a world where timelines are shifting outward. While earlier stage preclinical programs are seeing only minor delays to their plans, most of the clinical-stage programs will likely face more material impacts because the medical centers where trials are commonly run have been forced to pause elective treatments and clinical studies. Deciding how to adapt to this new environment is the key to success.

Like others in the early-stage biotech ecosystem, we are actively working with our executive teams around the strategic planning process,[2] to ensure we adjust to the new R&D execution reality and have the adequate funding to get to key data inflections. This often involves tradeoffs in the pipeline and buy-ups/downs across the budget—key elements to driving optionality and adapting to the new reality.

In addition, biotech start-up DNA also has a strong community gene: sharing insights in a noncompetitive, constructive manner. As an example, we responded to the crisis by helping bring biotech executives together and created a forum for best practice sharing[3] across the biotech industry. This forum shares insights around managing through this crisis (e.g. lab operations with physical distancing, remote engagement with investors, protective measures, etc.). These and similar efforts across the industry further help start-ups adapt to the new and challenging operating model.

Biotech's ability to adapt and learn quickly under different circumstances is a core strength of the start-up model, in particular during the real-time urgency of a pandemic.

Science guides us, now more than ever

At its essence, our early-stage investment philosophy takes a science-first approach to venture creation. We start with great science, whether it's from academic labs, spinouts from corporate partners, or born of the minds of entrepreneurs in our offices. We help construct a business model that fits the scientific thesis, with the mission of advancing a medicine to patients. And then build an organization to deliver that mission.

But it all starts with great science: strong rationale for a therapeutic hypothesis, robustly reproducible and well-designed experimental data sets, with a clear strategy for translating the science from bench

2. https://lifescivc.com/2020/03/strategic-planning-in-biotech-during-a-pandemic-crisis.

3. https://twitter.com/atlasventure/status/1252936177063211008.

to bedside for patients. These are the key ingredients for successful biotech enterprises.

Importantly, great science reinforces the primacy of data in our decision-making. In the words of W. Edwards Deming, "In God we trust; all others bring data." For the biotech ecosystem, as noted above, data is the primary currency by which we value progress. Colorful anecdotes of unproven but miraculous observations are generally not the domain of high-quality biotech companies, but rather of ephemeral elixirs and snake oil. And the plural of these kinds of anecdotes is not data. Sadly, during this pandemic we've seen plenty of examples of the latter.

While new therapies can and should be informed by personal narratives of patient impact, evidence-based medicine requires a high bar for data integrity and experimental design. In the clinical realm, for example, this involves conducting randomized and controlled clinical trials, preferably double-blinded with comparator arms to remove bias in observing the impact of any interventions.

The recent urgent push for drugs to treat or prevent COVID is a great illustration of the importance of well-designed clinical data: many early observations of "game-changer" drugs turned out to be illusory. Sadly, there's a long history of this in clinical practice. We need to resist the urge to jump on anecdotes by relentless focusing on gold standard data. Fortunately, the Food and Drug Administration, our principal regulatory body, has set a high bar for science, and plays a crucial role in continuing to uphold this high standard, which is good for patients, payers, and the biopharma industry.

Because of biotech's reliance on rigorous data and sound science, even in the face of pandemic, we remain bullish on the fundamentals of the sector: innovative discoveries, grounded by the primacy of data, will continue to be advanced into therapies for patients.

Further, now more than ever, it's clear that rigorous science will be what gets us out of this pandemic, with biopharma leading the way with new anti-COVID therapies for patients and society.

Bruce Booth is a partner at Atlas Venture, an early-stage biotech VC firm, where he invests in new therapeutics start-ups. He's had the privilege of creating or working with over two dozen biotechs over the past two decades. Bruce also enjoys writing about biotech in a widely syndicated blog, *LifeSciVC*.[4]

Atlas Venture is an early-stage biotech venture capital firm. With the goal of doing well by doing good, we have been building breakthrough biotech startups for over twenty-five years. We work side by side with exceptional scientists and entrepreneurs to translate high impact science into medicines for patients. Our seed-led venture creation strategy rigorously selects and focuses investment on the most compelling opportunities to build scalable businesses and realize value.

4. http://www.lifescivc.com.

SHOWING UP TOGETHER WHEN IT MATTERS

Alexander Karnal

At Deerfield, we put our $10 billion of capital to work behind those willing to challenge the status quo with a clear goal of advancing health care. Our team numbers almost 150 talented and kind people who come together to do something that matters every day.

Our work, whether through our upcoming life science hub, the Cure, or through the Institute for Life Changing Medicines, has never been more urgent. We are bringing together leading life science, digital health, and tech-enabled companies from around the world to develop innovative health solutions for the most pressing needs. In the end, we hope to deliver on our simple belief that a healthy life is our most basic human right.

On March 8, 2020, I awoke to a very different environment, one where COIVD-19 was turning the world upside down, financial markets were in freefall, and many people were left white-faced and shell-shocked. Not us. One thing I love about our team is that we have been together through many wild environments, and this time was no different. Together, we knew we would find our true north and charge forward.

Looking back, for all the warnings from public health experts over the years that it is not a matter of if, but when, before the next global pandemic, COVID-19 was hard to imagine. Despite our increasingly globalized economy and the fact that pandemics strike roughly once every hundred years—making us due for one—the possibility, let alone probability, did not feel real. Yet, here we are, and the global COVID-19 pandemic is a painful reminder of the critical role played by our country's leading innovative scientists, biotech companies, and investors.

When the pandemic hit, we quickly mobilized to face the new reality, working across our investment firm to understand how to get ahead of this fluid event, leveraging all analyzable details in real time, even as our lives were being personally impacted. In studying the volumes of insights coming in every minute, it quickly became clear that COVID-19 could spread rapidly and have lethal consequences. Even more daunting was how little was known versus theorized about COVID-19 and how complex it would be for the world and us to navigate through to the other side.

Recognizing the extraordinary and unprecedented threat that is COVID-19, we decided to take our unique insights and act on them. We rallied as a team and promptly developed a plan, first to protect people, but at the same time looking ahead to reopening the economy and getting back to normal—a new normal—and envisioning what that would look like. Our plan was grounded in asking difficult public health questions to inform rational decision-making for the public good.

A critical component of our plan was the capturing and analysis of data in a centrally organized and strategic way for informed decision-making. With no room or time for miscalculations on critical dimensions, we understood how crucial having this data was, given how little was known about the virus. As other companies scrambled to produce antibody tests, many of which turned out to be flawed, we marshaled our collective expertise to take a broader view and joined forces with an artificial intelligence company to develop a user-friendly data-tracking app. This comprehensive tool, which will monitor for recurrence of the virus and could also be used for biosurveillance, enables protective policies that will help guide decisions about back to work, back to school, and back to normal.

The app integrates a wellness monitoring function with the goal of early case identification and symptom tracking. It indicates when self-quarantine, PCR testing, and contact tracing by Apple and Google platforms are necessary. In addition to enabling the prediction of emerging hot spots, the data are expected to empower the anticipation of, and preparation for, the next potential wave. The app, which was launched for free because it was the right thing to do, was available in May 2020.

We are optimistic that this plan, driven by the collective and broad insights of a remarkable group of experts at Deerfield, will provide a path forward and help to answer many critical questions. Though it is not without its risks, it minimizes risk in an informed and data-driven manner. Broad testing, within the context of a multiphased, adaptive, and risk-tolerant approach, is in society's best interest, to be able to move forward. Without robust data, it is as if we are flying a plane without navigation.

On the back of all this Deerfield work, I was invited by Governor Ned Lamont to join his task force on reopening the state of Connecticut. Like the other states in the tristate area, Connecticut has been trying to manage COVID-19, and was among the first to release a framework for progressing from a stay-at-home order to the new normal. Deerfield had been engaging with Governor Lamont on this question from the outset of the local outbreak. As an author of the reopening plan, I committed my business skills, scientific understanding, and deep connectivity to our team and resources at Deerfield to help protect and reimagine the state of Connecticut.

Beyond the current need to don facemasks and institute a moratorium on normal life, our work helped to reveal the importance of identifying new models of investment in biomedical defense as we move forward. To protect against the potential use of biological weapons or the next pandemic, however remote, and contribute to this discussion, we came together to identify a number of innovative ways to incentivize investment in this vital area.

Here, our approach is grounded in lowering risk and improving the certainty of potential return. Getting ahead of and taking proactive steps to improve the odds of reaching these critical benchmarks may be necessary, given how unknowable the time frame is for when we need to leverage these urgent innovations to save lives. Details of this plan, which we expect could create a new paradigm that drives the substantial scale of for-profit financial and intellectual capital to eliminate any future risk that we may all face, have been presented.

As for therapeutics to treat COVID-19, ultimately, we anticipate having a range of options to lower the risk of life-threatening complications from the virus, marking a major development and great comeback for our country.

The critical lesson here is that we must always ensure that the US is a leader in innovation, never let down our guard, and always be one step ahead of the unimaginable. We must maintain an adaptable environment that allows investors to invest so that scientists can invent. This forms the basis for a virtuous cycle of innovation needed in order to yield an armamentarium of life-changing innovations, which are essential to not only our growth but our vitality.

Imagine for just a minute that we did not have the uncapped promise of financial return to bring the billions of dollars and millions of people together in pursuit of answers and solutions to life-threatening events. Rather than just planning for our new normal over the next several months, we could be scrambling to hide while bearing witness to and being a victim ourselves of the unthinkable. Fortunately, that is not the case and we have been given fair warning. COVID-19 is a reminder and a wake-up call of why that can and should never be.

While it might feel dark now, our nation has the resolve for a revival and will benefit from the learnings from this complex time, making us stronger than ever.

Alexander Karnal is partner and managing director at Deerfield. He was appointed by Governor Ned Lamont to the Reopen Connecticut COVID-19 Advisory Group. He also serves on the boards of the New York Academy of Medicine, Biotechnology Innovation Organization, the Institute for Life Changing Medicines, Recovery Centers of America, and Discovery Labs, the latter three for which he is a cofounder.

Deerfield is an investment firm dedicated to advancing health care through information, investment, and philanthropy—all toward the end goals of cures for disease, improved quality of life, and reduced cost of care. As of 2020, the firm manages nearly $10 billion in assets.

PERSPECTIVES OF A BIOTECH INVESTOR

Mark Lampert

I am not a pandemic expert, nor a scientist. Rather, my perspective comes from a long career investing in biotechnology companies. I have learned much from witnessing promising companies fail, and some risky ones succeed. I believe an appreciation of risk and investment is relevant to our response to COVID-19, and to preparation for future pandemics.

Investment Lesson No. 1: Seemingly rare, "black swan" events drive most outsized profits and losses. Outlier scenarios occur with surprisingly high frequency (because the spectrum of possibilities is so large), and they nearly always manifest in unforeseen ways. Only a few months ago, who imagined where we would find ourselves today? Project this thought forward by a year: the spectrum of heretofore unimagined possibilities is breathtaking. Human nature tends to underestimate the odds for the unfamiliar.

Investment Lesson No. 2: Beware of false bottoms. When bad news strikes a company, its stock price at first plunges and then often stabilizes at a lower level, the "new normal." We want desperately to believe that the worst is behind us (and sometimes it actually is), so we refrain from selling the stock. Unfortunately, bad news is often followed by more bad news. Our beloved stock plunges further, we are shocked anew, and the pattern continues until it doesn't. Pandemics are analogous. In the midst of highly uncertain, rapidly unfolding circumstances, we latch on to each piece of good news, hoping that the worst is over. The reality, however, is that going forward the likelihood of extreme outlier scenarios is greater than before the pandemic began. ("Problems can create their own momentum."—Warren Buffett.) We must be careful not to drop our guard after the first wave of bad news subsides. Outlier scenarios (both good and bad) lurk beneath the seemingly stable "new normal."

Investment Lesson No. 3: When we incur a significant investment loss, we hold a post-mortem to try to discern whether we made a prospective mistake. Sometimes the conclusion is that no mistakes were made, we made a good ex ante investment, but we got unlucky. Given the same facts and circumstances in the future, we would make the same investment decision again. Other times we acknowledge our ex ante mistakes and pledge not to repeat them. Perhaps society's lack of adequate preparation for COVID-19 was more the former case, because such a devastating pandemic was unprecedented in our lifetimes. However, we would be afforded no such excuse if we remain unprepared for the next pandemic.

Investment Lesson No. 4 (with credit to the renowned investor Howard Marks): The most profitable investments often involve two features: a) envisioning a future that is at odds with what most other people envision, and b) being right. Putting these two together is difficult and rare. Unorthodox predictions are usually wrong; those who espouse them are often viewed as quacks. (And many are!) Accordingly, the following are some thoughts related to pandemics that I believe fall outside mainstream discussion. You can judge the rest for yourself.

1. An investment in preparation for future pandemics may be the highest-return investment that society could ever make

As COVID-19 has starkly demonstrated, pandemics are not merely theoretical risks; they are extraordinarily costly, in both dollars and lives, and there is vast room for improvement in our preparation and response. The investment needed to achieve an order of magnitude improvement is insignificant relative to (a) the risk-adjusted economic value of the improvement, and (b) the investment required to achieve equivalent improvements with respect to other existential threats (e.g.

nuclear war, climate change, etc.). The world spends approximately $2 trillion per year on military defense, yet a relatively minuscule amount on defense against our common global pathogenic enemies. Throughout history, more people have died from pandemics than from war by a factor of ten. A rational visitor from planet Mars, upon learning of this discrepancy, might question the existence of intelligent life on Earth.

2. Humility

For hundreds of years, epidemics were assumed to be caused by "miasma," or foul air. For thousands of years, bloodletting with leeches was "known," with certainty, to be therapeutic. Widely accepted dogma can be dangerous and can drown out due consideration of alternative hypotheses. Often, it is the nonexpert outsider who successfully challenges conventional wisdom (i.e. Gregor Mendel, Billy Beane). Crafty pathogens, the beneficiaries of millions of years of evolution, will reveal their own truths and vulnerabilities over time; overconfidence could cause us to miss them. In the fog of war against fast-moving pathogens, a few erudite-sounding experts, whom we desperately want to believe, can find themselves suddenly empowered to make life and death decisions for our entire population. The best antidote is an a-politicized scientific inquiry that follows rapidly emerging data with openminded objectivity.

3. Risk segmentation

COVID-19 has variously affected several distinct population segments: the severely ill, the moderately ill, asymptomatic carriers, high-risk frontline workers, low-risk healthy individuals. The optimal interventional strategy is different for each segment. Greater risk is acceptable in managing and treating a severely ill patient, but less for a low-risk, healthy individual. One important implication is that drugs to treat the severely ill may be developed faster than a vaccine for the low-risk, healthy population because the safety thresholds, and requisite proof thereof, are vastly different. (I worry about prematurely injecting unprecedented forms of vaccines into billions of arms simultaneously.)

4. Serendipity

Luck has always played a critical but mysterious role in drug development. No matter how well we think we understand a disease and a drug candidate, the drug's safety and efficacy in patients are always uncertain until we run the experiment (most drug candidates fail). Nature made the human body far too complex for foolproof predictions. Serendipitous observations have led to the discovery of blockbuster drugs (e.g. Viagra, penicillin), while other "rationally designed" drugs have failed unexpectedly. These two approaches to drug discovery are, in fact, complementary. In a pandemic, serendipity can be maximized by diversifying the considered hypotheses and by increasing the interventional "shots on goal."

5. Preemptive therapeutic development

In the face of exponential pandemic growth curves, time is a critical parameter. The traditional drug development process was simply not designed for speed. Redesigning the process from scratch, to optimize for pandemics, will yield enormous benefits, specifically in compressing the time from pathogen identification to the unrestricted availability of lifesaving therapeutics for the severely ill. Traditional drug discovery, involving the random screening of small molecules to identify "hits," followed by "lead optimization" and biological testing, is too iterative and laborious for pandemic drug development. By the time an Ebola therapy was discovered, the epidemic was over. Creative alternative approaches must be identified, evaluated, and implemented well before the next pandemic strikes.

- Remdesivir, the first drug approved for COVID-19, was developed a decade ago for a different disease. Given what was already known about remdesivir from previous profiling, it was among the first agents to be thrown against COVID-19. The lesson: advanced characterization against known pathogens of thousands of existing drug candidates can greatly shorten the discovery timeline when the next pathogen emerges. Additionally, the speed of manufacturing scale-up should be an important factor in the prioritization of drug candidates. Finally, an alternative approach to pandemic drug discovery is to preemptively identify and assemble new platforms of therapeutic "building blocks" with the capacity to rapidly translate the genetic sequence of a pathogen into a readily manufactured designer drug. Creative minds will undoubtedly come up with other novel approaches, with speed being the unifying theme.

- Traditional drug development involves lengthy randomized, placebo-controlled clinical trial programs for every drug candidate (i.e. ten drug candidates require ten distinct clinical trial programs). In a pandemic, what is needed is the sports equivalent of a "two-minute offense": an unorthodox strategy designed specifically to address the extraordinary value of time. The situation is ideally suited for creatively designed "adaptive" clinical trials of numerous drug candidates run in parallel.

A question heard frequently these days about investing and life generally, is: "when will things stabilize and return to normal?" The truth is that the future is never predictable or stable. The very concepts of "stability" and "normalcy" are much more reflections of our collective psychology at a given moment than they are accurate descriptors of empiric truth. There always exists a wide range of "alternative histories," including both good and bad outlier scenarios. Our future selves may look back on this period as a much-needed wake-up call that has revealed the extent to which we were completely and unnecessarily exposed to pandemics. We hopefully will dodge an existential bullet this time as a fortuitous consequence of COVID-19's lethality rate, age discrimination, and slow propagation. In contrast, the next influenza pandemic, if at all like the one from 1918, could sweep the world in a matter of weeks, killing tens of millions of young and old indiscriminately. The best and only defense lies in serious and sophisticated preparation.

Mark Lampert is the founder and general partner of BVF Partners LP ("BVF"), a San Francisco-based private biotechnology investment firm established in 1993. Lampert has been active in the biotechnology industry since 1984 and has served on the boards of directors of numerous public and private companies. He is a founding donor to UnorthodoxPhilanthropy.org and to GiveDirectly.org. Born and raised in Ann Arbor, Michigan, Lampert holds an AB in Chemistry from Harvard College and an MBA from Harvard Business School.

BVF Partners L.P. is a San Francisco–based private investment partnership specializing in fundamentally-driven public biotechnology investments. Since its inception in 1993, BVF has strived to build concentrated, long-term investments in small-cap biotechnology companies while performing rigorous diligence and ongoing monitoring of its investments. BVF is committed to working with its portfolio companies as partners in their successes.

WHAT BIOTECH INVESTORS NEED FROM FEDERAL AGENCIES IN THE COVID CRISIS

Geoff Porges

Since the emergence of the pandemic spread of SARS-CoV-2 in Italy on March 6 and 7 (when the daily case number broke through 1,000), US equity investors have been deeply concerned about the effect of the COVID-19 pandemic on every sector of the economy. In the three days after the breakout in Italy, the equity markets in the US lost more than 20 percent of their value, and while they have partly recovered, since mid-March there have been more than thirty-four million new unemployment claims, and a full-blown depression is considered possible. Amid this disarray, biopharma stocks have been a relatively safe haven, and, after a few weeks of inertia, the industry has pivoted its global R&D might toward the threat posed by COVID-19.

Politicians, investors, and even the general public are realizing that the only way to save tens of thousands of lives, and recover millions of jobs, is from biopharmaceutical innovation. With support from government agencies, foundations, and traditional investors, capital is supporting major new R&D investments, and new research programs are emerging daily. These programs are being supported by new collaborations, multicompany partnerships, and academic initiatives. With all this private sector and academic activity, we continue to hear a refrain from investors and executives—what are the federal agencies doing?

Conventional wisdom holds that biopharma executives and investors would prefer that government agencies and regulators get out of the way, or at least do the minimum necessary, while industry develops tests, medicines, and vaccines on a loosely regulated basis. This is not the view we hear about the COVID situation from investors, nor from the leadership of the more established companies. Mostly they have a consistent plea regarding the agencies, that "we need them to do their job." We summarize here what these constituencies need and expect from key federal government institutions in this crisis.

The agencies of most importance to investors and biopharma executives are the National Institutes of Health (NIH), the Food and Drug Administration (FDA), the Centers for Disease Control (CDC) and the Centers for Medicare and Medicaid Services (CMS). All these agencies have tremendous influence over the direction of industry R&D, and also on the willingness of investors to support the industry and deploy capital to COVID and related indications. The market hates uncertainty and assumes the worst in a vacuum. When these agencies are silent, or, even worse, ineffective, it creates concern in the investment community.

NIH

As with the other agencies mentioned here, biopharma and biopharma investors count on the NIH for fundamental research into virology, biology, pathophysiology, immunology, and potential treatment targets. So far through COVID, the NIH has been relatively invisible, other than the canonical guidance of Dr. Anthony Fauci. Investors and industry executives are looking for this enormous research institution to "do its job." This means providing expansive and insightful research into the nature of the virus, its genome, its proteins, its structure, its behavior, its immunology, and the pathophysiology of the disease it causes, and distributing that information as soon as possible to the academic and industry community. The NIH should not endorse ill-founded medicines rushed to market, and should not speculate about drug and vaccine development timelines, product profiles, and disease impact. The NIH has so far lagged its Chinese counterparts in providing important research about COVID to the scientific and medical community, and in this shortfall has left industry and investors looking to experts in Asia and other areas for prepublication research and commentary about the virus's characteristics and the treatment of the disease. Regrettably, the premier research institution in the US is in danger of losing its stature and reputation in the fight against the most disruptive medical threat of a lifetime.

FDA

The perception among investors is that the FDA has swung from excessive caution regarding COVID to excessive flexibility and accommodation. Investors perceive that there have been too many Emergency Use Authorization approvals, and see a lack of clarity about development standards and hurdles for new tests, drugs, and vaccines. They do not regard this as in the interests of investors, the industry, or consumers. Clear guardrails are the least investors expect, and as pundits and presidential advisers have suggested, the first priority is addressing the virus, in a rigorous and scientifically responsible way.

Investors are suggesting that they remain confused about the reliability and interpretation of polymerise chain reaction (PCR) and antibody tests, and see the FDA as being slow to regulate or manage speculation about off-label use of medicines for COVID. From an investor's point of view, the commercial introduction of unproven products crowds out the development opportunity for products that are attempting more rigorous development. It steals investment capital, and gives the impression of closing out market opportunities for more transformative and thoroughly studied products. Investors look to the FDA to maintain stringent oversight of safety, efficacy, and product integrity in all approvals, so that patients and their families and providers are not put at risk, or led astray, by flimsy assessments, limited data, or false claims. Accelerated development is a worthy goal, but not at the price of patient safety or clarity of information and utility. There is a long history of politically influenced accelerated development of medicines and vaccines in the face of epidemic and pandemic threats. Many such approvals have caused significant harm (polio, influenza vaccines, AZT).

More than seventy COVID vaccine development programs are underway, almost all addressing the same consensus S protein target. Investors are hopeful that the FDA will establish standards for how clinical efficacy will be demonstrated in different populations, what safety requirements need to be met, what surrogate markers are acceptable, and how they should be measured. As in diagnostic development, vaccine development is perceived to be a free-for-all in need of guidance. Investors are being told different things by different companies, and at least one vaccine developer and the federal government have communicated substantially different timelines about vaccine development.

Once again, the investors who supply the capital for these large development efforts want the FDA to set up the goalposts, decide the dimensions of the field, mark the goal lines, and then referee the game fairly according to those standards. Once those rules for COVID vaccine and therapeutic development are established, investors will have much more confidence in the development plans and timelines being proposed.

CDC

The CDC was once considered the preeminent public health agency in the world, with experts in epidemiology, virology, bacteriology, statistics, and many other disciplines. The agency contributed substantially to eliminating smallpox, and near elimination of poliomyelitis and measles, to name just a handful of their many achievements. But one of the biggest questions about the pandemic response (and one sure to be examined in post-hoc assessments), is the lack of visibility and effectiveness of the CDC in response to COVID-19. The agency had a window of time to respond to the threat ahead of its emergence (January and February), and is generally regarded as having fallen short of its responsibilities during that period. Looking ahead, the CDC should be setting the standards for testing rates and ratios, for contact tracing procedures and policies, for educating and deploying a flexible corps of thousands of contact tracing personnel, and for setting the guidelines about the procedures and precautions for businesses and public organizations as they begin to reopen after COVID. While the CDC is not traditionally responsible for vaccine development, it does manage the National Immunization Program, and is responsible for the influential Advisory Committee on Immunization Practices that shapes national immunization recommendations. To build the confidence of investors, provide accurate information about future vaccine needs and priorities, and establish the ground rules for widespread use of COVID vaccines, these constituencies look to the CDC to be "doing its job," providing expert epidemiological and public health leadership, guidance, and oversight.

CMS

Medicaid and Medicare are likely to be the largest payers for the diagnostic, therapeutic, and vaccine products for COVID that emerge in the coming months and years. Remarkably, of these four federal agencies and institutes, CMS has been the most proactive and effective in its response to COVID. CMS has already established clear reimbursement policies for COVID-related transportation, hospitalization, and diagnostic and therapeutic interventions. CMS has made it clear that it will pay for the costs of COVID care for beneficiaries of Medicare, and in mid-February introduced a new HCPS billing code for laboratory testing or SARS-CoV-2. At this time, CMS had the foresight to create separate codes for PCR and other COVID tests at any laboratory or health care facility. The agency has declared its intention to cover COVID vaccines, and explicitly confirmed its coverage for hospital services for COVID. The agency authorized reimbursement for outpatient services related to COVID, including prescription drugs, and approved reimbursement for telehealth services. All these policies are positive for the industry, and for investors. At the very least, it clarifies that tests, medicines, products, and vaccines that are developed for this disease will be covered for a large proportion of the affected population. However, even CMS will need to adapt its reimbursement policies to cover the cost of testing and surveillance of asymptomatic individuals for public health reasons, rather than medical ones.

Overall then, the message from the investment community regarding federal government agencies and centers in this pandemic is simple—after three months of relatively modest progress, and frequent missteps, the community of investors in biopharmaceutical companies would prefer these agencies and institutes to fulfill their traditional responsibilities, to the best of their abilities. This means maintaining the high standards of professionalism, rigor, integrity, transparency, and insight they have met in the past. If they don't serve these constituencies, and more importantly the American public, in this way, then when we collectively survey the wreckage of this pandemic, in a few months, or a few years, they may find themselves buried in it. Biopharma needs sensible, rigorous transparent advice and guidance, funding, reimbursement, and yes, regulation. The best way these agencies can facilitate the development of the transformative tests, drugs, and vaccines that can alleviate and then end this pandemic is by simply doing their jobs.

Dr. Geoffrey Porges is the director of therapeutics research and a senior research analyst at SVB Leerink covering diversified biopharmaceutical stocks. He brings to the firm over twenty-five years of expertise in advisory, executive, and investment roles in the biopharmaceutical industry.

Dr. Porges earned his medical degree from the University of Sydney and trained in pediatric and internal medicine in Australia. He is also a graduate of Harvard Business School, where he was a Baker Scholar and was formerly responsible for the commercialization of vaccines in the vaccine division of Merck & Co.

SVB Leerink is a leading investment bank, specializing in healthcare and life sciences. The firm's knowledge, experience, and focus enable it to help its clients define and achieve their strategic, capital markets, and investment objectives.

LOOKING TO THE FUTURE

THE BIOTECH SOCIAL CONTRACT THROUGH THE COVID LENS

Peter Kolchinsky

**Graphics included here are excerpted from RA Capital's COVID-19 TechAtlas Map. For the full map of the biopharmaceutical industry's response to COVID-19, visit www.racap.com/covid-19.*

In the age of COVID, the nonessential has been stripped away, revealing who and what is essential. Not only do we express our gratitude to the nurses and doctors on the front lines of the pandemic, but our appreciation for teachers, supermarket cashiers, cleaners, and delivery workers has deepened.

When COVID struck, the public called on the biotechnology industry to innovate, to develop tests, treatments, and vaccines as quickly as possible. And the industry responded, with hundreds of companies pivoting to develop any countermeasure they could to mitigate the crisis. Diagnostics companies worked on tests to detect COVID infections. Biotech and pharmaceutical companies studied existing drugs and even failed R&D programs to see if anything on our shelves could quickly help save lives. Many companies also started research programs from scratch to find and target the new virus's particular vulnerabilities. Vaccine manufacturers began retooling their technologies to inoculate billions of people against a new pathogen faster than ever before. Nonprofits and governments around the world recognized the need to help the industry fund R&D and invest in manufacturing capacity. Squeezing what is typically a decade-or-more-long gantlet of R&D, regulatory review, manufacturing, and distribution into a single year requires ingenuity, perseverance, unprecedented investment, and coordination.

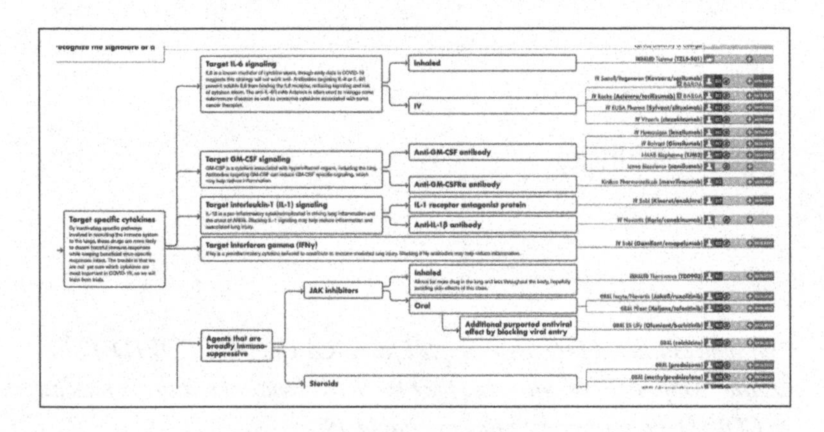

Meanwhile, insurance companies did what they so often do. Although their spending on health care plummeted as patients avoided hospitals and put off all but the most urgent procedures, insurers, robotically tone-deaf to the crisis, continued to reject whatever costs their bureaucracies could find excuses to reject. Early on, some plans refused to cover costs for COVID testing and treatment. While that is exactly what has been happening for many years to patients with cancer, diabetes, and countless other conditions, COVID drew attention.

COVID puts the structural flaws of the American health care system on full display. For the uninsured and underinsured with health problems before COVID, those faults were already visible. But now, our uniquely American inability to provide adequate health insurance to our population threatens to inflame an already dire public health and financial tragedy.

Elected officials and the media called for insurance plans to fully reimburse COVID testing and care, and those plans responded. The federal government vowed to pay COVID-related medical bills for anyone without insurance. In other words, when it came to COVID, America suddenly and finally saw the need for insurance reform, called for it, and made it happen.

With millions of Americans losing jobs and consequently their health insurance, some drug companies redoubled their efforts to help patients with out-of-pocket costs. For example, Eli Lilly guaranteed that no patient would pay more than $35 a month for insulin, necessary assistance that was long overdue (and yet wouldn't be necessary if America had a functional insurance system). The drug industry organization PhRMA started promoting MAT.org, a website that matches patients struggling with out-of-pocket costs with patient assistance programs for the drugs they need. That, too, was long overdue. COVID has accelerated all kinds of reforms.

The pandemic is teaching us what is truly essential. Medicines are essential. And when we are threatened by a disease for which we don't have treatments, then maintaining a vigorous industry, academic, and government response to develop new therapies and tests is essential.

Today we are talking about COVID. But it is essential for all of us to remember that cancer, diabetes, Parkinson, and many other diseases are personal COVID-level crises for those who suffer from them. The innovation and effort now directed at COVID has long been directed and will continue to be directed at all those other diseases, provided we recognize their importance and value.

Essential knowledge

Information, clear thinking, and level-headed analysis are also essential. In response to this emergency, my team at RA Capital did what it has always done: research, teach, and provide funding. We assembled a COVID response team to evaluate and map out all drugs and vaccines that are being developed to fight or prevent the disease, a chessboard to visualize how these pieces might be combined and sequenced in a coordinated strategy against COVID. Our work is normally proprietary and shared only with companies we invest in, but we released our COVID map to the world. Our insights from the map led us to connect companies and laboratories to link technologies or assays that made sense together. We held calls with diagnostics and vaccine companies to share knowledge that the sides weren't yet exchanging. We offered our knowledge and services to the Massachusetts Emergency Task Force on Coronavirus. And we helped a couple of our companies pivot to developing drugs against the virus.

We also studied the importance of social distancing, realizing it would be required longer than most people appreciated, with punishing consequences for the global economy. For many of our biotech companies, social distancing, especially from hospital hot zones, meant extensive delays in clinical trial enrollment. We knew early on that these companies would need help funding their operations for a longer stretch before they could generate the data that would enable them to raise more money. We consulted and collaborated with like-minded biotech investors to continue to fund companies to get them through a longer period of uncertainty. Unfortunately, making finite resources stretch further meant shutting down certain research projects, perhaps to be restarted down the road.

We are still at the beginning of our fight against COVID, but already there are positive signs. As of late May, we are tracking dozens of vaccines against SARS-CoV-2, many of which are already being tested in humans. It's likely that our industry will break the record for vaccine development speed—by a matter of years—and have a COVID vaccine available at least for frontline workers and the most vulnerable by the end of 2020.

In the search for treatments, dozens of drugs are being tested in clinical studies to shorten the duration of infection or reduce its lethality. This first wave of drugs that reached clinical testing emerged from our vast medicine cabinet, which includes many inexpensive generic drugs, currently branded drugs, and even some still in development. Some were already financial successes; others have been investment sinkholes, having failed in development for their original purpose. The only drug that has so far shown positive results in a proper clinical study in COVID patients, Gilead Sciences' remdesivir, was originally developed to fight Ebola.

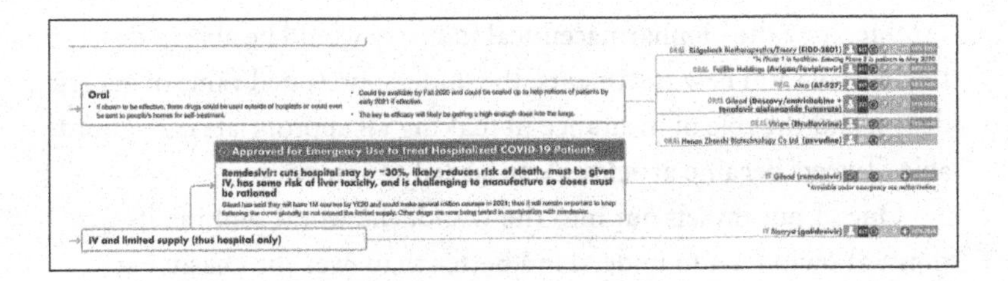

RA Capital TechAtlas COVID-19 map: excerpt of map of drugs targeting cytokines to treat cytokine storm in severely ill patients

Before COVID, "repurposing" was a dirty word. We were conditioned to tout novelty because that's what the public expected of us. Now everyone just wants results. If those results come from a drug originally used for rheumatoid arthritis or HIV, who cares? After COVID, we will hopefully still see the ingenuity and prudence in extracting all the value we can from all the drugs we already have.

Of course, we're unlikely to find a perfect COVID cure among them, so discovering entirely new drugs remains essential. In the next wave of trials, we'll test drugs designed specifically to target SARS-CoV-2. Yet even those discovery tools and the experience of all the scientists now working on COVID were repurposed from other therapeutic areas. We must ours ourselves appreciate and help the public and legislators appreciate the power of incremental progress.

Distributing unbiased public health information is also essential. The public is hungry for clear explanations for why this virus is so tricky, how we might beat it with vaccines and treatments, and what we can do to stay safe until those therapies become available. Though it's uncommon for investors to directly engage with the public, I believe we should spread our knowledge as much as possible, especially in a crisis. Turning to the modern public square of Twitter and dusting off my virologist credentials, I have tried to do my part by writing science explainers. Advanced concepts become analogies—antibodies are police dogs; viruses, evil robots; their genomes, pamphlets in a copy machine. In 280-character bites, I engaged with the American and global public on the importance of social distancing, how SARS-CoV-2 infects us, how it's changing but not changing fast enough to elude a vaccine, how the biopharmaceutical industry would be able to develop that vaccine and new treatments, the underappreciated value of generic drugs, and the role of insurance in making all appropriate care affordable. Tweets became articles. Articles led to interviews.

Once I put myself out into the world, the world sent me its questions, allowing me to understand better than ever the chasm between innovators and the people for whom they innovate. It wasn't hard to see how we had managed to do much to advance public health but failed to inspire the public. To some of them, our industry is a cost, a drain on society that needs to be reined in by Congress. But hundreds of people also respond with messages of hope, appreciation of science, scientists, and even the drug industry, and interest in learning more.

Innovation as an essential, affordable resource

Like COVID testing and hospital care, our eventual drugs and vaccines against COVID should be accessible to everyone, so we can lift our world out of the pandemic. The end goal of every innovator is not just to invent a treatment that could theoretically address an unmet need, but to have that drug be accessible and affordable to patients so it can actually meet their need.

The outcry over drug affordability often results in calls for drug price controls that would squelch the kind of important innovation that has given us a head start on fighting COVID. Now, amid this crisis, America has seen how essential proper insurance is for public health and patched some of the gaps in America's coverage to ensure that COVID tests and treatments are fully covered.

Maybe we can hold onto to these insurance patches and turn them into real reforms. Maybe drug companies will continue to expand their patient assistance programs that cap out-of-pocket costs not only for those without health insurance but for anyone regardless of income or insurance status, recognizing that trying to weed out the few uninsured patients wealthy enough to afford branded drugs creates obstacles to millions who can't and struggle with paperwork to prove it. And maybe we'll remain open to recognizing the value of finding new uses for generic drugs and continue to fund that research. I hope so. And I hope the biopharmaceutical industry will live up to the brief glow it will enjoy as saviors of the world. I hope it will shun (and accept that Congress must reform) the rent-seeking business model of unduly delaying the genericization of their drugs, decoupling revenues from innovation. Because if there's anything that COVID is teaching us, it's that our industry is at its best when driven to innovate. Innovation is essential. It is and will always be our safety net.

When people shout "Make COVID testing free!" or "A COVID vaccine should be free for everyone!" while also calling for "More funding for COVID vaccines and treatments!" what they are really clamoring for is continued investment in innovation and a health care system in which everyone is properly insured and receives the care their physicians prescribe without punitive out-of-pocket costs. What is true for COVID treatment is true for cancer treatment, too. It's true for diabetes, rheumatoid arthritis, HIV, and genetic diseases. How we solve COVID—with continued innovation and insurance reform—is how we can solve every other illness that threatens us and our children. It's long past time we shouted for the right reforms.

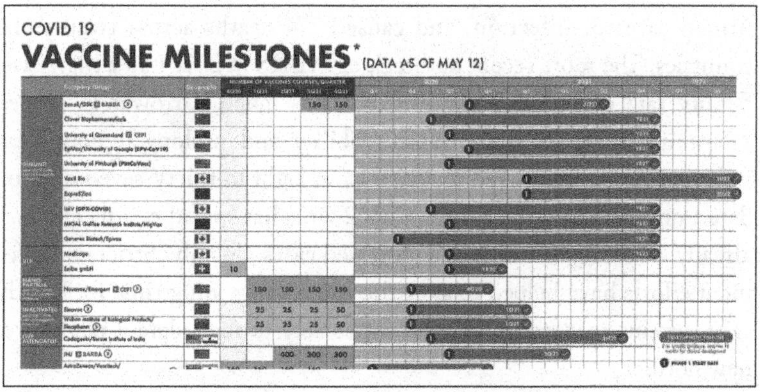

COVID 19
VACCINE MILESTONES* (DATA AS OF MAY 12)

Company/Group	Number of vaccines courses/quarter: 4Q20	1Q21	2Q21	3Q21	4Q21
Sanofi/GSK BARDA				150	150
Clover Biopharmaceuticals					
University of Queensland CEPI					
EpiVax/University of Georgia (EPV-CoV19)					
University of Pittsburgh (PittCoVacc)					
Vaxil Bio					
ExpreS2ion					
IMV (DPX-COVID)					
MIGAL Galilee Research Institute/MigVax					
Generex Biotech/Epivax					
Medicago					
Saiba gmbH	10				
Novavax/Emergent CEPI		150	150	150	150
Sinovac		25	25	25	50
Wuhan Institute of Biological Products/Sinopharm		25	25	25	50
Codagenix/Serum Institute of India					
JNJ BARDA			400	300	300
AstraZeneca/Vaxitech/ [illegible]	[illegible]	[illegible]	[illegible]	[illegible]	[illegible]

Peter Kolchinsky, PhD, a biotechnology investor and virologist, is managing partner of RA Capital Management LP, which builds and invests in companies that develop drugs, diagnostics, and medical devices. He is also the author of *The Great American Drug Deal: A New Prescription for Innovative and Affordable Medicines*.

WHY 'NORMAL' WILL BE DIFFERENT FROM NOW ON

Yaron Werber

The COVID-19 pandemic represents the third and most severe outbreak of coronavirus since the turn of the century. The 2002–03 spread of SARS lasted about seven months, included more than 8,000 confirmed cases of infection, and caused 774 deaths across twenty-six countries. The more recent outbreak of MERS in 2012 had a high case fatality rate of about 35 percent, but was much narrower in scope. Stemming from its high transmissibility and prolonged period of infectiousness—estimated to range from eight to thirty-seven days in data from Wuhan—the SARS-CoV-2 virus has spread rapidly and has already surpassed 3.3 million confirmed cases globally. Strict containment efforts have helped to slow the spread, providing time for health care systems to decompress and for society to formulate a plan for a new normal.

The initially exponential growth curves have demonstrated encouraging signs of flattening thanks to social distancing and other measures (Figure 1). These efforts appear to have brought the effective reproduction rate (Re) below 1 in many parts of the world, though we cannot know conclusively, given the high number of unconfirmed cases. If we assume the Re is now below 1, then maintaining strict social dis-

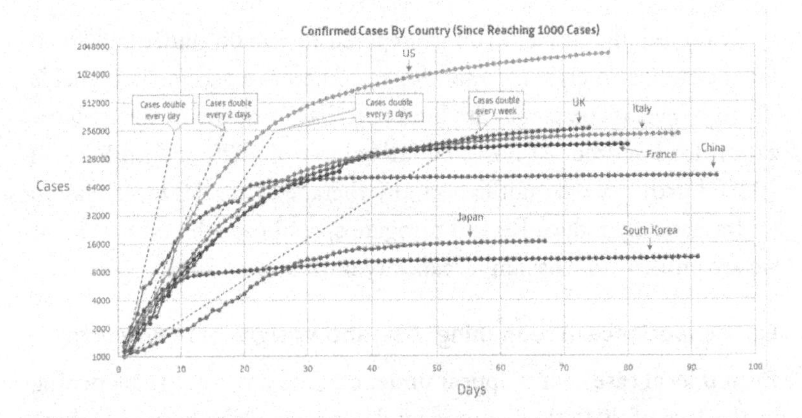

Figure 1 Case doubling time by country shows slowing spread with containment efforts (as of May 26, 2020)[1]

tancing indefinitely (until a proven vaccine emerges) will eventually extinguish the virus. But given the high economic cost of an indefinite shutdown, parts of the world have already started to reopen non-essential activity despite the risk of a second wave. In the US, every state began at least a partial reopening by mid-May, following much of the EU's reopening gradually in April.

Data indicates resurgence is likely if containment efforts are eased

Work from Harvard's School of Public Health has projected that a second wave (with others thereafter) is unavoidable without indefinite social distancing until a vaccine is available. If social distancing is eased before a vaccine, as we are now seeing in many places, their model shows widespread infection (more than 50 percent of the population) within a short time; the rapidity depends on whether social distancing is intermittently reinstated or abandoned. The contagion would stop only when herd immunity is reached.

1. Johns Hopkins CSSE, ourworldindata.org, Cowen & Company.

It is clear that society is not prepared to remain under lockdown for the twelve-to eighteen-month period before a vaccine is available (though several companies are hoping to have emergency use authorization for hundreds of doses by fall/winter of 2020 and millions in 2021). Further, we do not expect any therapeutic under investigation to dramatically reduce Re and mitigate spread over the near term, as we are unlikely to have ample capacity until at least 2021.

Early experience in reopening: Asia shows signs of resurgence

Even if local case counts appear under control, any country reopening in the face of 80,000 or more new daily cases globally risks looking imprudent in hindsight. Cases are still accelerating in Latin American countries, such as Brazil, Chile, and Peru, and new clusters are reported in eastern Europe, the Middle East, India, and other regions. Asia is seeing new outbreaks from travelers and spread in crowded dorms of migrant workers. Hong Kong, Singapore, and Taiwan are implementing new restrictions, and Japan has now implemented emergency plans. Singapore was initially heralded as a model of COVID-19 response but overlooked clusters of cases among migrant laborers led to a rapid resurgence.

As Europe gradually reopened, several countries (Spain and Germany, for example) saw resurgences leading to extending some restrictions. Outbreaks are expected to hit in waves in different regions of the world as containment is eased, leading the WHO to recommended countries not to lift restrictions too quickly.

Summer weather is unlikely to obviate the need to prevent a second wave

Past pandemics have consistently shown a substantial burn off in transmission/infection during warmer, more humid summer months. These are almost invariably followed by a second, even larger wave in the following fall/winter.

Recent studies from the National Biodefense Analysis and Countermeasures Center suggest that SARS-CoV-2 on surfaces is susceptible to increased heat, humidity, and especially direct sunlight, while the impact on airborne or aerosolized virus has not been clearly established. However, public health officials and epidemiologists have warned that the lack of herd immunity will likely overpower any benefit from this summer heat, should public health measures like social distancing and self-isolation end. This notion is bolstered by outbreaks in much warmer climates, including Australia, Iran, Brazil, the Middle East, and Singapore.

Current population immunity is likely far from herd immunity levels

If we will need 50 to 60 percent (or more) of the population to be resistant to achieve herd immunity, screening studies are the best way to estimate disease prevalence and learn how close we are. To this point, molecular testing has been geared to those with symptoms, given limited capacity, and thus total confirmed case count is a vast underestimate of true cases. Serologic testing holds the key to defining the proportion of the population that has been exposed to SARS-CoV-2, with the caveat that work still needs to be done to determine the reliability of immunity in individuals with antibodies present. Our review of the durability of immunity with other coronavirus infections suggests that reinfections are possible (six of nine patients were reinfected within a year), although whether this is extrapolatable to SARS-CoV-2 is unknown.

Serologic testing was performed as a screening tool in Santa Clara County, California, and found a population prevalence of 2.5–4.2 percent. While this is still far from herd immunity levels, it was fifty to eighty-five times as high as the number of confirmed cases at the time. Random testing done by New York State of 3,000 people at grocery stores and shopping locations found antibodies in 13.9 percent of those tested, with a higher rate among those in New York City at 21.2 percent and about 18 percent in Long Island; it was 3.6 percent in upstate New York. This result is in line with a previously performed sampling of pregnant women in New York City, which showed an antibody prevalence of 15 percent. The highest reported prevalence was in a Boston homeless shelter population, which indicated 36 percent of residents had acquired COVID-19, though this is a particularly vulnerable population and does not represent the likely prevalence in the region.

Testing done by Amgen/DeCode Genetics in Iceland showed herd immunity rates of 0.6–0.8 percent since the outbreak, and a study in Gangelt, Germany, showed a 15 percent rate in the hardest-hit area of that country. In a recent study performed in the UK, antibodies were detected in 17 percent of Londoners and 5 percent of the rest of the country.

Testing will be key during reopening

Integral in the White House plan for reopening is sufficient testing so that we can immediately identify new cases and contact trace. As of mid-May, thirteen million tests had been completed in the US (thirty-seven million globally), with over 300,000 tests being done per day. Cowen's life science and diagnostic tools team expects molecular testing capacity to increase to 500,000 to one million per day by June. The team expects 100 million molecular tests by 2021, 50 percent in the US. Additionally, capacity for serological testing should reach fifty million to 100 million by June.

However, a bigger challenge has been organizing care and access to tests. In addition, due to different testing criteria, there are wide geographical variations in the percentage of positive tests, ranging from 2 percent in South Korea to 12 percent in the US and 21 percent in the UK (Figure 2).

Gradually relaxing social distancing with strict measures

Italy, Germany, France, Austria, Spain, the Czech Republic, Finland, Denmark, Norway, and other countries relaxed restrictions by mid-May to reopen some business and allow some outdoor activities to take place. China lifted the eleven-week lockdown in Wuhan on April 8. Some countries, including China, Germany, Iran, South Korea, Lebanon, and Saudi Arabia were forced to reimpose lockdown measures due to case resurgence. The US is following, with major states showing signs of stabilization, including heavily affected states such as New York, New Jersey, California, Michigan, Florida, and Wash-

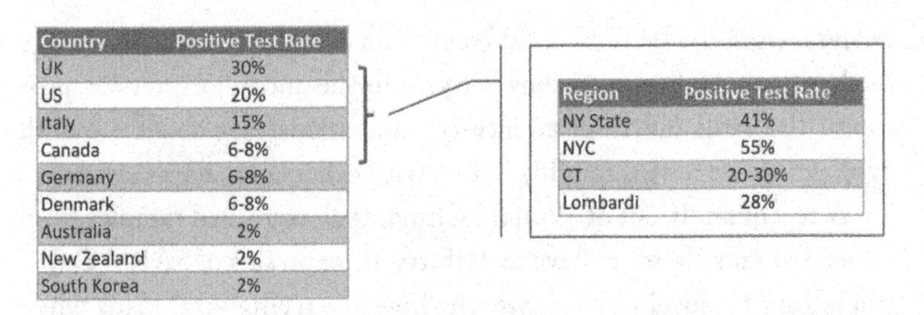

Country	Positive Test Rate
UK	30%
US	20%
Italy	15%
Canada	6-8%
Germany	6-8%
Denmark	6-8%
Australia	2%
New Zealand	2%
South Korea	2%

Region	Positive Test Rate
NY State	41%
NYC	55%
CT	20-30%
Lombardi	28%

Figure 2: Wide variability in positive testing rates globally likely a consequence of capacity constraints and different testing criteria[2]

ington. Most states are yet to see a significant rebound in cases in the early stages of partial reopening.

The new "new normal" will be highly variable from country to country and state to state and might be driven by testing availability and hospital capacity. We believe governments will lift the lockdown while implementing some new measures, such as gradually reopening of shops and businesses with visitor restrictions, implementing body temperature checks, and relaxing social distancing requirements. Large events will remain off the table for the foreseeable future. High-risk populations (such as the elderly and individuals with comorbidities) will remain isolated, and low-risk populations will likely proceed cautiously. Putting it all together, we believe the economic recovery will be closer to a gradual return to pre-COVID-19 economic activity than a "V" shape.

More so, the new new normal will be highly variable depending on location, occupation, and time of year. This will require flexibility, adaptability, and individual commitment to the overall welfare of our communities.

Mutagenesis is unlikely to outpace vaccine development

Biostatistical analyses of COVID-19's mutagenic profile are ongoing, and predictive conclusions from these studies shift frequently; however, they remain essential to the development of a viable vaccine. Collective

2. Cowen and Company and *The Atlantic*, April 16, 2020.

analyses from the Bedford lab at Nextstrain currently project an average of about twenty-five mutations per year in the antigen expression profile of the virus (up from twenty-two mutations/year about a month ago), which we note is roughly in line with other coronavirus strains.

A recent study out of Wuhan, China, that examined samples from eleven patients claims to have seen thirty-three strains of SARS-CoV-2 (including 19 novel strains), which showed varying viral loads when tested in vitro. Importantly, the study had not yet been peer-reviewed, though it also suggests that select mutations in the virus may confer additional pathogenicity. Broad sequencing in Iceland detected five strains originating from the UK and mainland Europe.

Recent work from the Los Alamos National Laboratory (not yet peer-reviewed) identified fourteen novel mutations in the spike protein of SARS-CoV-2, including the D614G mutation in the coronavirus spike protein, which has grown into the dominant strain of the virus around the world. While this mutation appears to significantly boost the transmissibility of the coronavirus, there does not appear to be any significant difference in pathogenicity or mortality associated with the mutation. Continued screening/documentation of novel mutations in COVID-19 will be essential as therapies targeting these proteins advance through development.

Although the findings have yet to be widely corroborated, SARS-CoV-2's mutation rate suggests that each successive vaccine will need to account for future variants of the virus in the same way that flu vaccines attempt to address the emergence of new variants each season. However, the rate of key surface antigens targetable by a vaccine still appears considerably slower than the flu and is not expected to drastically shift within each season.

The role of CD4+ and CD8+ T cells in COVID-19 has recently been characterized in studies out of Germany (the Charité University Hospital in Berlin) and the US (La Jolla Institute for Immunology) by examining the serum of recovered patients. Identifying the natural immune response to the virus can help in vaccine strategy and in the selection of immunological endpoints for vaccine trials.

The German and US studies found CD4+ T cells reactive to the SARS-CoV-2 spike glycoprotein in 83 percent and 100 percent of COVID-19 patients, respectively. Beyond the spike protein, the US study found T cell recognition of the M and N proteins, which are likely codominant as each was recognized in 100 percent of patients, and several additional structural antigens. Taken together, this data identifies many antigens that can potentially be included in a vaccine toward the goal of mimicking the natural T cell response after infection.

Vaccines and neutralizing antibodies are what will expedite economic recovery

Positive data for drugs like Gilead's remdesivir have led to broad market responses, which may decrease the length of hospitalization and modestly improve need for oxygen therapy, have led to broad market responses. Advancements of this nature are certainly good for hospitalized patients who are moderately (pending Phase 3 data) or severely (positive Phase 3 data has already been released at the time of this publication) ill but are unlikely to "bend the curve" that is needed to fully reopen the economy and improve consumer confidence.

We need safe and effective vaccines or therapeutics that can be used prophylactically to prevent spread, and effective oral therapies that would be used upon diagnosis in the outpatient setting to prevent spread and reduce hospitalizations. We lack good oral anti-COVID-19 antivirals in development.

Once a patient is hospitalized, COVID-19 can cause a cytokine storm and coagulopathy that leads to pneumonia, acute respiratory distress syndrome, acute renal failure, myocarditis, and other cardiac complications. Other therapeutic candidates that fit this category of potentially helping severe hospitalized patients but are not helpful for preventing spread or "bending the curve" would be IL-6 antagonists (for example, Roche's Actemra and Sanofi/Regeneron's Kevzara), convalescent plasma, complement inhibitors (such as Alexion's Soliris or Ultomiris), JAK inhibitors (such as Pfizer's Xeljanz, Lilly's Olumiant,

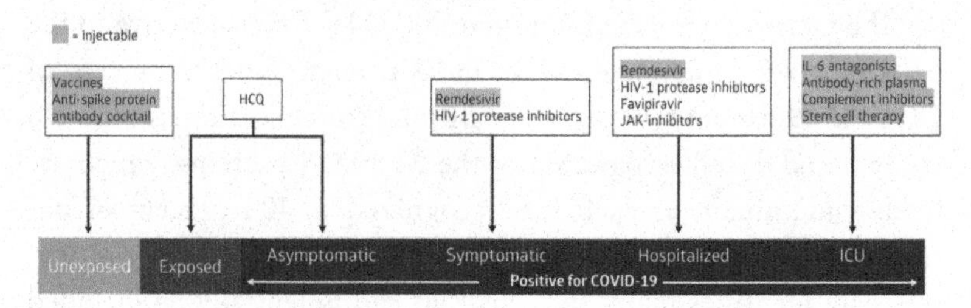

Figure 3. Administration period for therapeutics in testing for COVID-19[3]

and AbbVie's Rinvoq) and anticoagulants. These therapies are all given after a patient is hospitalized, often in critical condition (Figure 3).

In contrast, some potential developments would be game changers for economic recovery in the face of COVID-19, namely vaccines, prophylactic agents that decrease risk of infection (data for HCQ as post-exposure prophylaxis is expected in the coming weeks), and anti-spike protein, neutralizing antibody cocktails (such as Regeneron, Vir, and Amgen).

One small step backward, one giant leap for humanity

The COVID-19 pandemic abruptly disrupted our everyday lives, our freedoms, our sense of stability, and the economic fabrics that makes our societies flow. The epidemic touched each of us in a different way and reaffirmed the importance that social energy, community, emotional and professional connectivity, and sense of belonging, and productivity bring to being healthy. To many, this has been a period of self-growth, reflection, refocusing energy on new ideas and strategies, or finding time to reequilibrate and reconnect with immediate families. To many, the time at home has been enjoyable, whimsical, and challenging; to some stressful and isolating. Yet to so many, this epidemic tested their ability to cope, stretched their financial reserves, and shook their stability.

While this too shall pass, we will all be forever transformed by this historic period and our kids will tell their grandchildren about the time

3. Cowen and Company.

they first used Zoom. While our pets have been the major beneficiaries, we would argue that the ability to persevere, innovate, refocus our sense of purpose and community, and care for one another will ultimately turn this public health tragedy into many yet unforeseen leap forward to us as people.

Dr. Yaron Werber is managing director and senior research analyst on Cowen's biotechnology team. Previously, during his 20-year career, he was a managing director and head of US health care and biotech equity research at Citigroup. Before rejoining Cowen, Yaron was a founding team member, chief business and financial officer, treasurer, and secretary of Ovid Therapeutics. He began his career at SG Cowen as a biotech analyst after receiving a combined MD/MBA from Tufts.

A HOPEFUL LETTER FROM THE FUTURE

Jeff Kindler

If I had a grandchild born in March 2020, I hope I'd be able to write this letter as their high school graduation approached in the spring of 2038.

The world has been on a remarkable journey since you joined it.

As I'm sure you're tired of hearing, everything changed in profound and unexpected ways the month you were born. Suddenly, stunningly, the entire globe shut down.

We were isolated in our homes and could interact with family, friends, and coworkers only through very primitive 2D/5G technology. We lived in a surreal and dystopian nightscape—experiencing tedium and terror, stress and anxiety, depression, and, sometimes, hope.

Those of us privileged to possess good shelter, food, companionship, and technology watched helplessly as tens of millions lost their jobs, businesses that they and their families had spent lifetimes building, and even their homes and their access to food.

And the sickness, the pain, the suffering was almost too much to bear. The deaths . . . so many deaths; ceaseless stories of painful deaths; young, old, healthy, sick; thousands of deaths every single day all over the planet.

Our health care system seemed woefully inadequate to the catastrophe we faced.

You've grown up in a world of scientific and technological advances that have so dramatically reduced illness, improved health, and extended life, that it's probably impossible for you to imagine what it was like the year you were born. Challenges that must seem to you easy to overcome seemed insuperable to us.

We have come so far. So, how did we get from there to here?

Even before the Crisis hit, our health care system was inequitable, inefficient, and overly complex. Far too many received poor quality care—or no care at all. Too many others couldn't afford it. Obtaining and paying for treatments required providers and patients to navigate fragmented, opaque, and painfully complicated arrangements.

Hard as it may be for you to conceive, many parts of the health care system still relied on technology from the 1980s—and, believe it or not, records were often on paper. Too often, the patient or their advocate had to supply providers with their medical history, as best as they could remember it. Can you even imagine your generation—who grew up with the ability to summon everything from wines to cars to a dinner date from your personal device—tolerating that?

While we fitfully took on discrete issues around cost, access, and quality, we lacked the will and imagination to improve the health care system in comprehensive, durable, and meaningful ways. As a society, we had many competing priorities presenting equally formidable demands for reform—education, infrastructure, the environment. And, as with those matters, political partisanship paralyzed policy making.

The biopharmaceutical industry—private companies, large and small—employed tens of thousands of scientists, researchers, and other professionals who dedicated their careers to discovering and developing new treatments. They made essential contributions to the improvement of health and human life.

Yet, even as they continued to generate extraordinary medical innovations, these firms were demonized by the media, politicians, and the public. Sometimes the actions of some people in those firms warranted contempt, but more often the hostility derived from a belief that the work lacked value commensurate with its risks and rewards.

But the Crisis changed all that.

Imagine waking up one day and suddenly facing a deadly and highly contagious novel virus with no vaccines, no therapies, insufficient protective equipment for caregivers, no means of quickly producing and scaling diagnostics and antibody tests, no efficient systems for gathering, analyzing, and reporting real-time global data, and an ever-changing set of uncoordinated and scattershot government responses.

As broken, fragmented, and inefficient as our health care system was BC, it came completely unglued in the Crisis. We struggled with overburdened and inadequately supplied hospitals, overworked health care providers facing unbearable stress and personal risk, and minimal if any care for people with acute and dire illnesses that didn't make the triage cut.

Even medical events that usually produced joy were suffused with sadness. My heart broke when your father was told he couldn't be with your mother when you were born.

In other words, the Crisis reminded us in the starkest possible ways that nothing else matters if we don't prioritize investments aimed at preventing sickness, treating disease, and prolonging life.

The Crisis inspired a renewed reverence for the human spirit unlike any I've seen in my lifetime. The arts found new ways of inspiring and lifting us; technology found new ways of binding us; colleges and high schools found new ways of educating us; existing and new businesses found new ways of reaching us; communities found new ways of celebrating their heroes and themselves. Having seen clear skies, we newly cherished the environment.

Those of us with healthy, loving families treasured them more than ever. Social distancing reminded us of the incomparable worth of a simple hug. And those of us with family and friends who suffered or died prized life and good health more than ever.

And so, we urgently resolved to address the many challenges and disparities in our health care system. The harsh failures of that system, vividly exposed during the Crisis, combined with an energetic pursuit of innovation to drive meaningful and lasting change.

We accelerated long-overdue technology-enabled improvements in access, quality, and delivery of care. Hospitals and providers adopted new, efficient, and more sustainable business models. Public and private payers worked closely with government, academic, and private researchers and manufacturers. Together, they found better ways of incentivizing innovation while broadening access to the miraculous output of the world's laboratories.

And, regarding the virus itself, we witnessed unprecedented global cooperation of scientists from the public and private sectors, massive financial investments, real-time innovations in discovery and development, pragmatic speed, and flexibility from regulators.

Within a few months of the outbreak, medical researchers around the world had developed more than 250 diagnostic tests and were running clinical trials for more than 100 treatment candidates and at least 15 potential vaccines.

Together we ended the Crisis—and better prepared ourselves for the next one—through collaboration and competition. That established the model for bringing forward the many additional medical miracles that have so benefited your generation and those to follow.

Along the way, the public experienced a new appreciation for the work of the biopharmaceutical industry. At the height of the Crisis, Henry Olsen wrote in *The Washington Post*: "The world's much maligned pharmaceutical industry is expected to come to the rescue at breakneck speed. Let's treat them as the heroes they are when they ultimately deliver."

Of course, those of us involved in that industry knew that our colleagues' achievements during the Crisis built on a long legacy.

Ours has always been an enterprise dedicated to improving lives through science and it has always risen to the challenges of the moment. That's been our heritage from Heinrich Merck's scientific industrialization in the early 1800s, to Edward Squibb's focus on medicinal quality during the 1846 Mexican-American War, to Colonel Eli Lilly's institutionalizing private research and development after the Civil War, to Pfizer's factories saving thousands of soldiers' lives in World War II by scaling Sir Alexander Fleming's extraordinary discovery.

And by the time the Crisis hit, our industry had produced many more game-changing advances in treatments and cures for hundreds of different diseases. The death rate from HIV/AIDS had dropped nearly 85 percent since peaking in 1995. In the prior decade, cardiovascular death rates dropped by nearly a third. By 2020, there had been a 22 percent drop in cancer death rates since their peak in 1991. And, of course, that was just the beginning of the amazing treatments and cures developed during your lifetime.

All this and the enormous further progress in the nearly two decades since the Crisis, make it possible for me—an eighty-three-year-old in better health than I was at the time of the Crisis—to sit in the front row of your high school graduation this spring. As I watch you and your classmates walk across the stage, I'll be thinking about the medical and technological advances that your generation will produce in the ongoing project of improving people's lives.

But I will also be thinking about my pride in the biopharmaceutical industry—whose people always have and always will play a unique and powerful role in that life-affirming project. As they had so often before, when things looked very bleak, those people came through again.

For all the pain and suffering of the Crisis, I think it can be said, to borrow from Churchill, that, for the biopharmaceutical industry, this was our finest hour.

Jeff Kindler is CEO of Centrexion Therapeutics, a biotech developing nonaddictive pain treatments; operating partner at ARTIS Ventures, a venture capital firm; chair of GLG Institute, an executive learning community; and a director at several public and private health care companies. Previously, he was chair and CEO of Pfizer.

RETHINK HOW WE DO THINGS: CREATE AND SHARE KNOWLEDGE TO WIN BACK THE FUTURE

Otello Stampacchia

As I write, in late April 2020, we are a couple of months into a virtual lockdown in Boston. We are among the fortunate ones: we still have jobs and good salaries, can work from home and provide for our families, and are connected to an "essential" industry (biomedical research and development) by being investors in health care start-ups.

But, perhaps because of that, I personally feel a huge burden of responsibility. To my family, here and in Italy. To my colleagues. To the companies that we have created and invested in, who are all trying to assist however they can.

It is now finally and painfully evident to everybody that this pandemic has and will continue to have massive consequences for everybody's life and our collective futures. Personal consequences, with virtually everybody having lost dear ones or living in constant fear for their safety. Economic consequences, with tens of millions of unemployed in the US and many, many millions more across the world; entire industries going toward extinction; supply chains being virtually redesigned from scratch; and unprecedented fiscal deficits to ensure people can weather the necessary containment measures. And there will be many more, far-reaching consequences that we still cannot predict: we are still shell-shocked by how "this" has happened and what it means to even think about the "then."

You know, we have been lucky, relatively speaking. This will not be a comforting or popular message as lives are being lost or ravaged by this pandemic. Think of a similar virus, with roughly 50 percent of infected people being asymptomatic and about two weeks from infection to hospitalization for severe patients. Now increase the fatality rate from about 2 percent to about 10 percent, like the first SARS coronavirus that emerged in 2002–2003. It would have probably wiped out our civilization as we know it. And this will happen again. There are estimates that bats alone have hundreds of thousands of viruses that we have not been exposed to, like this coronavirus. There are many more, and quite possibly more dangerous viruses in swine, birds, and other animal reservoirs that we know nothing about. On a planet now teeming with eight billion humans, traveling and interacting constantly, it is only a question of time before this happens again.

As an industry, and as a society, we need to figure out a way out of this, and make sure it *never* happens again. How to do that is something society needs to start carefully thinking about.

We had been forewarned. We did not listen. For decades, virologists and epidemiologists, among many others, have been raising the alarm about the emergence of viruses with pandemic potential, from multiple places across the world. For decades we have had regular outbreaks of such epidemics, which in most cases have been contained more or less "locally": SARS, MERS, H5N1, Zika, Ebola, Marburg, etc. Bill Gates famously gave a TED Talk on this topic in 2015.

The fact that we did not better prepare, as well as our lacking response in the early phases of the pandemic, will be recorded as a collective failure of global leadership that probably rivals or exceeds the mishaps that caused the great wars of the twentieth century.

How can we respond and lessen the damage being inflicted, and at the same time make sure it is never repeated? Start by embracing the scientific method and heeding the opinion of experts in the field.

Wikipedia says the scientific method is "an empirical method of acquiring knowledge that has characterized the development of science since at least the seventeenth century. It involves careful observation, applying rigorous skepticism about what is observed, given that cognitive assumptions can distort how one interprets the observation. It involves formulating hypotheses, via induction, based on such observations; experimental and measurement-based testing of deductions drawn from the hypotheses; and refinement (or elimination) of the hypotheses based on the experimental findings."

The same methods need to be brought to bear to solve this crisis and prevent the next one. A necessary part of the solution is the biotechnology and pharmaceutical industry, which, as soon as the pandemic hit, shifted resources to try to solve the pandemic. More than 800,000 people work in the US in the biopharmaceutical industry, which directly or indirectly supports 4.7 million jobs across the country. As one, these people are now trying to find a cure, a vaccine, or drugs that could lessen the infection's severe symptoms, as well as designing and creating more effective diagnostics.

Many more thousands of scientists and researchers, in universities and hospitals across the world, have stopped all their other research to study this novel virus and how it wreaks havoc in patients' bodies. Thousands of research papers are being published instantly, eliminating the archaic, long, inefficient journal publication system, and subject to immediate analysis and peer review by thousands of other scientists in real time.

We are likely to find out more about our immune system and how it interacts with infectious agents (as well as cancer and other disease states) due to this crisis than we did discover in the past one hundred years. The dividends of such a coordinated and massive research effort will propel science and medicine at a velocity unforeseeable until four months ago. This approach in information sharing needs to remain a cornerstone of scientific research after we go back to "normal," whatever normal will mean at the end of this pandemic.

Science and the scientific method will, in time, find a solution to this. It will never be fast enough, but we will get there. As I said above, they are a necessary ingredient to the solution. They are not sufficient. We need to radically rethink how scientific research is prioritized and funded. We also need to think about our societal and decision-making processes.

In today's world, emerging infectious threats spread fast. It is tempting to pick the "medieval option," raise the drawbridges and let everybody fend for themselves. Unless you really are considering going back to the Dark Ages, that will not work. Even then, when it took literally months to go from Asia to Europe, ships were kept away from ports, often for forty days ("quarantine" is related to the Italian *quaranta*, forty), and with a total world's population of only 300 to 400 million people, pandemics routinely wiped out entire villages and decimated countries. Smallpox, rabies, plague, measles, and many others, probably killed more than 500 million people over the last couple of centuries. And that was without airplanes, cruise ships, trains, cars, bikes, and other faster means of transporting illness.

One thing will always travel faster than the virus: information. We just need to gather it, analyze it, and share it—with the scientific method as a guide, and not filter it through political or religious lenses—and avail ourselves of that advantage. It is the only way to fight these threats: as a species, together, instead of as a collection of small communities. United, we stand. Divided, we will fall.

For decades, scientific research, and investment in education, have dwindled across the globe. The US had a 2018 military budget of roughly $694 billion USD. That represented about 36 percent of the total global military expenditure that year. The National Institutes of Health's total budget for 2018 was about $27 billion, just 3.89 percent of the US military budget. This pandemic, quite possibly, will cost the global economy north of $10 trillion.

Emerging, new viral threats are much more dangerous enemies for the human species than any other potential enemies, with the possible exception of climate change (by the way, the two are connected.) We need to dramatically increase funding for research to analyze and hopefully prevent these threats. We need to sequence every single virus currently in the known animal reservoirs, identify potential antiviral molecules against them, stockpile them. While we are at it, let's not forget other infectious disease agents, like drug-resistant bacteria and stop neglecting investing in those threats. And let's throw in some mandatory educational training in statistics, epidemiology, and contact tracing for the new generations (like the Boy Scouts of America: always be prepared).

This funding needs to be removed from the discretionary, politically seasonal allocation requests that are part of the current political process. There has to be a supranational organization, which is funded independently of yearly government budgets, to avoid political influence on what need to be extremely rapid, scientific-based assessments and decisions. Unfortunately, politicians are not equipped to respond to these threats in the time required to prevent great loss of life. Perfection is the enemy of the good when it comes to responding to pandemics: reaction speed is your friend.

My most fervent wish is that we use this epochal tragedy to improve who we are as a species. We need to "win back the future" for the current generation of young citizens, who will likely suffer the most from the economic damage being inflicted by our necessary containment response to this pandemic. This can happen by investing in sustainable, science-based infrastructure driven by research: clean energy delivered by efficient grids, elimination of polluting energy and transportation methods to reverse climate change, better and cheaper medicines for diseases, and many others. The world rebuilt and invested after World War II: this ushered an era of unparalleled prosperity and lifted living standards for many.

We need to work together toward a similar future. The virus does not care about skin color, borders, or religions. Information and knowledge do impact and limit its spread and can provide us with the tools to defeat it. Science has been trying to warn us and it will find a cure, of that I am certain. Let's learn that important lesson once and for all, and let's not forget it when this is over. Because I fear we will pay an even bigger price in the future if we do not heed that lesson now.

Otello Stampacchia founded Omega Funds in 2004, one of the leading health care venture investment firms, with operations in the US and Europe. He is an Italian citizen and now lives in Boston, Omega's headquarters. He has written several pieces about the pandemic in the Timmerman Report: those opinions can be found on his LinkedIn page.[1]

Otello leads Omega's IR and strategic initiatives, is a member of the investment committee and is heavily involved in a number of Omega's therapeutic areas of interest, particularly oncology, infectious diseases, and immune disorders. He has been a director of a number of biotech companies, many of which have been acquired or listed on public stock exchanges. Otello has a PhD in Molecular Biology (from the University of Geneva, Switzerland), was one of the first recipients of the European Doctorate in Biotechnology (awarded from the European Union), and M.Sc. in Genetics (from the University of Pavia, Italy).

1. https://www.linkedin.com/in/otello-stampacchia-82aa5a19.

MUSIC AND MEANING

Paul Sekhri

I have been working in biotechnology for my entire career, but I have never been prouder to be affiliated with the life sciences field than I am now. We are all being tested by the coronavirus pandemic, in so many ways, and feel that those devoted to this industry have never been called on by humanity more than we are now. Each of us will respond to this crisis in different ways. And I believe that each of us will use our background and experience to do what we can to ameliorate this pandemic and focus our efforts on bringing it under control. As the CEO of a biotechnology company, my overriding responsibility is to ensure that our associates are safe while allowing them to remain as productive and connected as possible.

My grandfather on my mother's side was a physician. He was born and raised in Glasgow, but ended up living in London after graduating from medical school in Edinburgh. He was a family physician and had his "surgery," as they were called in England, at home, a stately Edwardian on a tree-lined wide street in Kilburn. I never knew him. His reputation was of a dearly loved family doctor who would make house calls day and night, someone who cared deeply about his patients, who loved classical music just as deeply. In fact, the story goes that he was headed to the opera (*The Bartered Bride* by Smetana) with my grandmother on a rare day off, when he pulled the car over complaining of chest pains. Asking my grandmother to duck into a nearby pub for a shot of brandy, he leaned over to her as she exited the car, kissing her and telling her he loved her. By the time she returned with the brandy, he was dead. I never knew him, but I felt like I did. His surgery remained untouched for years, my grandmother unable or unwilling to part with any of his books, medical equipment, manuscripts, and recordings; all sat there as if in a diorama. I spent many happy days poring through medical textbooks and musical scores, and was even given his stethoscope and slide rule when I intimated at an early age that I might want to follow in his footsteps.

He passed his love of classical music to my mother who, in turn, passed it to me and my brother. I started playing violin at the age of five, but quickly became enamored with the piano and studied it seriously for many years. It was not until I went to college that I finally decided that the pursuit of professional music performance might be a better avocation than vocation. But in so many ways, I have been repeatedly able to draw on the indelible imprint that music made on me.

Perhaps it should not be a surprise to find so many parallels between music and medicine. Nor does it come as a surprise (to me at least) at how many medical professionals either played or play an instrument, some exceedingly well. Music connects us. I see this when attending a concert in New York, at a recital at Carnegie Hall, watching the New York Philharmonic at David Geffen Hall, or attending the opening night of a new production at the Metropolitan Opera. What we experience as the music washes over us, seated in sold-out halls (and soon to be newly created logistics for attending these gatherings as a result of the pandemic), is intensely personal. At the same time, there are commonalities we all share. There is nothing in which I take greater pleasure than sharing thoughts and perceptions with friends and colleagues just after attending a concert together. I observe the conductor of a major symphony orchestra and it strikes me just how similar their job is to the CEO of a biotech company. The conductor and renowned speaker on leadership Benjamin Zander said, "The conductor of an orchestra doesn't make a sound. He depends, for his power, on his ability to make other people powerful." And the famous Estonian conductor Paavo Jarvi once said, "Conducting is like playing an instrument, but your instrument is people."

As the pandemic took hold and we were all relegated to our homes for quarantine and sequestering, I contemplated a way to maintain a vital connection with our employees, and thought what better way to do so than through my background in music. I am fortunate to call a few world-class musicians dear friends, and I reached out to one, Alisa Weilerstein, a world-renowned cellist, and proposed a series of online videos of her performing Bach. Each evening for two weeks, Alisanot only provided an incredible performance to our staff, but also began each video with a short discussion of what she, as a Type 1 diabetic, was dealing with. Her gorgeous music connected us. Perhaps that is a critical connection, one needed now more than ever.

While I did not become a physician or a concert artist, I did pursue a career in the biotechnology field. Over the years, I have held positions in tools and instrumentation companies, a small consulting firm focused on life sciences, big pharma (several times), small biotech (I was founding CEO of one biotech company then CEO of another), big private equity, and finally back to my real love, operating a biotech company.

I am CEO of a company with a singular and audacious mission. I am, moreover, determined that this mission will continue unabated in the face of, and despite, COVID-19. The goal of eGenesis is to eliminate the current waiting list for transplantable organs. Currently, in the US, well over 100,000 patients are on the transplant waiting list; 20 patients die every day waiting for an organ. eGenesis focuses on using cutting-edge gene editing and other technologies to modify the genomes of pigs, ultimately providing an unlimited supply of kidneys, islet cells (for Type 1 diabetic patients), livers, hearts, and lungs, eliminating the waiting list of patients who are in desperate need of a transplant. The current coronavirus pandemic has underscored just how critical it is what we, in the life sciences, do in terms of ameliorating disease, and in areas where medical need is so profound.

And research continues, albeit not at the pace it had been—first, our top concern is for the health and safety of our employees. Labs, which previously accommodated twenty to thirty individuals at any one time, have been emptied, with, at most, one or two researchers able to cohabit a lab at any given time. And even then, new social distancing recommendations (requirements?) dictate a careful intralaboratory ballet ensuring that no two individuals get too close. The silver lining behind this necessary isolation, and the need for most scientists to work from home, is, interestingly, a new ability to create thoughtful new hypotheses, ones that can and will be tested, verified, or modified in the lab once we can carefully reopen our facilities.

Since the start of my career, I have stated (to whoever would listen) that "people do business with people; companies don't do business with companies." In fact, my entire philosophy for living comes down to two words—*only connect*. When I was in high school, I read the novel *Howards End*, by one of my favorite authors, E.M. Forster, for the first time. On the opening page of the novel Forster writes these two words, "Only Connect . . ."

And much later in the book, Forster explains:

"Only connect! . . . Only connect the prose and the passion, and both will be exalted, and human love will be seen at its height. Live in fragments no longer. Only connect, and the beast and the monk, robbed of the isolation that is life to either, will die."

Forster seemingly longs for people to be able to really communicate one-on-one, in a significant and meaningful manner.

This novel was to have more impact on my life; my cousins were involved with the financing of the Merchant Ivory film, and I was lucky enough to attend its premier at the Cannes Film Festival in 1992 and walk into the Palais des Festival on the red carpet with the director and cast.

This idea of "only connect," for me, has been a powerful reminder about what is truly important in one's career, and in one's life. Perhaps there is no better time for us to focus on true and honest connections with others than at this time of isolation and social distancing. After all, if people truly do business with other people, the coronavirus pandemic has drawn into sharp focus what is superfluous and what is profoundly vital.

And one day, hopefully soon, life will return to a feeling of normalcy. While it is unlikely that things will return to exactly the way they were, perhaps we can find some optimism in this as well.

And, meanwhile, in my dreams at night I am still meeting with friends and colleagues; hugging them closely one-on-one at the opera, the concert hall, the theater. May this once again, and soon, become reality.

Only connect . . .

Paul Sekhri is the president and CEO of eGenesis Inc. Before eGenesis he held senior executive positions in small biotech (Cerimon, Lycera), private equity (TPG), and large pharma (Sanofi, Teva, Novartis). He has been a director on more than thirty private and public company boards, and is currently a member of the board of directors of Ipsen SA, Veeva Systems Inc., and Alpine Immune Sciences Inc. Sekhri is also the chair of the Board of Compugen Ltd., Pharming NV, and Petra Pharma.

Sekhri has a passion for cooking and classical music and is on the boards of several NYC-based institutions, including the Metropolitan Opera, the Knights, the Decoda Ensemble, and the Orchestra of St. Luke's. He also established the Life Sciences Council of Carnegie Hall.

REWRITING THE RULES OF HEALTHCARE

Wainwright Fishburn

As global chair of Digital Health at Cooley, I have worked with life science company-builders over decades. I was in the room at H&Q when Genentech's Bob Swanson presented the blockbuster/clot-buster TPA. Cooley served as founding counsel for the first major biotech companies. Today, Cooley is a new economy firm of more than 1,100 lawyers driving technological innovation around the globe. As a billion-dollar enterprise, we represent more than 50 percent of the NASDAQ Biotechnology Index and rank No. 1 by volume for biotech and pharma venture financings over the last decade.

Folklore has it that no one likes lawyers until you need one. Similarly, the life science/pharma industries have been maligned for drug pricing and a myriad of other criticisms. But now, more than ever, we need the life science industry!

Our industry is yielding solutions daily from billions of dollars of risk capital. Company collaboration around COVID-19 is beyond description, and the pandemic has accelerated the digitization of health care. Digital health—the convergence of life science/digital technology and health care—is one of the last sectors to be truly technology-enabled. Over the past decade, investors have poured $600 billion of private capital into life science companies (PitchBook) and more than $40 billion into digital health (Rock Health). Private investment capital, together with foundation and government funds, provide the resources available to fight the pandemic.

One silver lining is the convergence of innovators across industries working to beat the virus. Technology companies outside our industry have pivoted to help. For example, SpaceX is building critical ventilator components to help medical device-maker Medtronic fortify the number of these devices.[1] Imagine artificial intelligence and machine learning applied to PPE supply chain management. Clear AI, a UK-based company, maps real-time commerce between buyers, sellers, and supply chain providers. They are creating knowledge-based connective digital tissue enabling materially improved visibility, control, market access, and cost savings across medical supply chains. More than 275 UK-based health care organizations are engaged with this AI-based inventory forecasting technology as part of a joint initiative to track system-wide PPE for the NHS.[2]

Telehealth and remote monitoring are taking center stage as socially distanced patients are using telehealth visits before seeking in-person care (this was my personal story). Companies are heeding the call. Using biometric sensors connected to hospitals, providers, and first responders, Cherish Health is rolling out a home monitoring system for COVID-19 patients. Lumeon digitally screens and enrolls symptomatic patients in an SMS home monitoring program for NYC Health + Hospitals. On March 20, the US government announced that health care providers could practice across state lines, unlocking a greater supply of providers and eliminating some of the regulatory barriers that hinder early-stage digital health innovators from servicing clients nationally.

This shift toward modern telemedicine, in development for decades, is driving adoption of digital therapeutics and clinically/agency-validated,

---

1. https://www.cnbc.com/video/2020/04/06/medtronic-ceo-omar-ishrak-on-ramping-up-ventilator-production.html.

2. https://integratedcarejournal.com/newsdit-article/4750197c472b707014f2d1ed6418ed8e.

evidence-based software therapies. An iconic example is the connected blood glucose monitor allowing patients to check glucose levels and receive personalized feedback and recommended lifestyle changes. Digital pharmacies, such as Capsule, which offers same-day medication delivery to consumers, have also gained traction. Many companies have shifted efforts to combat COVID-19 for the good of humanity.

More than 150 of our clients are focused on COVID-19 solutions ranging from drug development to new technology platforms and applications. Helix, a population genomics company seeking to advance genomic research and accelerate the integration of genomic data into clinical care, plans to implement COVID-19 testing and carry out research for COVID-19-positive patients. Tasso, a member of Cedars-Sinai Medical Center's third accelerator class, is a blood-collection start-up working with the Department of Defense to conduct antibody testing. Vir Biotechnology and GSK entered a collaboration early in 2020 to discover, develop, and commercialize three programs directed to coronaviruses. Inovio Pharmaceuticals, which is developing a vaccine to prevent COVID-19, has received the FDA's approval to test the vaccine candidate in healthy volunteers. Celularity, a New Jersey-based cell therapeutics company, is investigating its proprietary cell therapy as a potential treatment option for the SARS-CoV-2 virus. This is the first known cell therapy to be granted an investigational new drug application by the FDA to treat COVID-19. On the technology side, Cigna and Express Scripts are working with AI health assistant Buoy Health to provide an early intervention screening tool to help consumers understand personal risk of exposure. Tech giants have joined forces to focus on contact tracing to fight the pandemic.

Other companies are relentlessly charging ahead. BrightInsight, a medical IoT platform for biopharma and medtech companies, announced a strategic partnership with AstraZeneca amid the early stages of the outbreak. AstraZeneca will develop apps, algorithms, Software as a Medical Device, and connected devices using BrightInsight's platform to enhance health care efficiencies for patients and providers. BrightInsight considered withholding the announcement of the AZ partnership given the cultural focus on COVID-19. In reality, the important message is that innovators such as BrightInsight continue to drive the economy forward despite the current state of the world. Likewise, we continue to see robust activity with venture financings closing during this time. The deep commitment to health care investment in "the best of times and the worst of times" remains steadfast.

This too shall pass, though the road ahead is far from clear. Let's continue to support the life science industry, as we need each other deeply—collaboration and innovation are paramount to our success.

Wainwright Fishburn, a prominent venture capital lawyer, is global chair of Cooley's Digital Health group. Nature recognized Fishburn as instrumental to San Diego's life sciences cluster. He serves on UCSD's Moores Cancer Center executive committee and is vice chair of Critical Path Institute. A cofounder of seven companies, including two public, he's also an avid cycler, recently circling the Cape of South Africa.

Clients partner with **Cooley** on transformative deals, complex IP and regulatory matters, and high-stakes litigation, where innovation meets the law. Cooley has 1,100+ lawyers across sixteen offices in the United States, Asia, and Europe.

BIOTECHNOLOGY AFTER COVID-19

Steven H. Holtzman

> There is one and only one social responsibility of business—to use its resources and engage in activities designed to increase its profits so long as it stays within the rules of the game . . .
> —CAPITALISM AND FREEDOM, *Milton Friedman*

> . . . a world in which science flourishes but justice is absent is condemned to the same fate as Sodom.—MURDEROUS SCIENCE, *Benno Mueller-Hill*

Looking ahead to the subsidence of COVID-19 as an existential threat, some social commentators have remarked on the need to reconceptualize what is or should be "the new normal." The new normal they envisage does not merely encompass designer face masks and elbow bumps instead of handshakes. Rather, it should be a "normal" that takes account of the social issues and divides—indeed, in a deep sense, social contradictions—that COVID-19 has laid bare. With respect to the United States, these include, in no particular order:

1. In the country's time of peril, those called upon to assume the greatest risk to their health and well-being for the benefit of the rest of the citizenry have largely numbered among the least well-compensated, the most disenfranchised, and the most

likely to lack health insurance, paid sick leave, and affordable child care.

2. When solidarity was most needed, when facts and (scientific) expertise were at their most paramount, and when consistent and honest, transparent leadership was most critical, the president and his enablers have continued to foster an environment of divisiveness while retailing anti-science, patently false, misleading, and contradictory narratives.
3. Our best hope for therapeutics and vaccines derive from our country's leadership in science—a leadership dually founded in the great democratic experiment embarked upon by the Founders less than 250 years ago and our country's embrace of immigrants regardless of their economic status upon arrival. But as we rely on that hope, an entire political party, along with its appointees to the Supreme Court, have issued policies and judgments that curtail electoral participation by the very same group of the most disenfranchised called upon to sacrifice most on behalf of us all, while promulgating regulations to limit the flow of immigrants, allowing in only those with preexisting economic means.

The behavior of the biopharmaceutical industry in the face of COVID-19 described in other essays in this volume stands in stark contrast. Old competitive postures were set aside in favor of solidarity and cooperative endeavor; protracted, legalistic contracting was often replaced by handshake (or at least elbow bump) agreements relying on the honor of industry colleagues; the best ideas were sought and embraced, regardless of the national origins, gender, sexual orientation, religion, political party, or ideology of their originators; and, facts and data, not economic and/or political power, determined the way forward. As several authors comment here, these behaviors represent our industry at its best; in the words of Lincoln's Second Inaugural Address, COVID-19 called forth "the better angels of our nature."

Against that backdrop, what does it mean for the biopharmaceutical industry to "return to normal"? As many of the authors here suggest, manifold lessons can be drawn for "our new normal" from our industry's response to COVID-19, e.g. open sharing of precompetitive information; cooperation among companies, NGOs, philanthropic foundations, and government agencies; risk-sharing to enable investments "ahead of the curve" in infrastructure and capabilities essential to the protection and well-being of our society as a whole; and a willingness to sacrifice sacred cows and to speak truth to power in favor of responding with agility to emerging factually grounded data, whether positive or negative.

In addition, some commentators have noted that, like conquering heroes returning victoriously from mortal combat, our industry has a shining moment in which to reestablish our good standing with the public: our industry was viewed among the most admired in the 1970s and 1980s, but had fallen to rank among the most vilified by December 2019

Yet, I find myself wanting to ask: is that all that should characterize the new normal for biotechnology? While no doubt salutary, the new behaviors can be readily embraced and supported in a Friedmanesque framework that stipulates that the social responsibility of corporate leadership begins and ends with the obligation to maximize profits. To do so, one need only extend from quarter to quarter the horizon in which the profits are to be maximized. From this perspective, the new normal is governed by the same principles as the old normal. The new behaviors and governing norms are embraced *contingently*, to be cast aside as soon as they are judged not to be expedient to profit maximization.

As for the industry's political engagement, the old rules will also continue to apply. Politicians, political parties, and social policies are to be supported (or not) based solely on their contribution to profit maximization (prototypically, support for corporate tax cuts and lower marginal income and capital gains tax rates, for surely the latter are necessary to attract the best, most capable to the leadership of our companies). Finally, seizing upon our newly restored reputation will be seen to involve little more than dispensing free doses of our new COVID-19 vaccines and therapeutics to people lacking health insurance, plus a well-designed and funded PR campaign.

In contrast to this perspective, I want to suggest that the biotechnology industry's new normal should be characterized by an engagement in the social/economic/political arena that is powered by and requires active advocacy for leaders and polices true to the fundamental founding principles and motivations of our industry. (Lincoln's call, after all, was not to seek extrinsic angels that might be of expedient utility; he entreated us to find those better angels in *our own nature.*) Conversely, our politics should involve active opposition to those who advocate policies that are inimical to our industry's nature. A politics of expediency may gain us a seat at the table; however, the price of a seat at the table is too high and too fraught if its cost is the loss of the moral authority of biotechnology's leaders to speak for our patients, our employees, and our communities.[1]

For me, our industry's essential nature and the political engagement it requires of us can be found in three areas: the industry's origins in the Enlightenment, the industry's commitment to human health, and the personal backgrounds, moral character, and motivations of key founders of our industry.

Enlightenment roots

Elsewhere[2], I have argued that biotechnology's roots in a commitment to the scientific method bequeathed to us by the Enlightenment is intrinsically connected to a commitment to the quest for social justice, the Enlightenment's second crowning bequest to the modern

1. For a further discussion, occasioned by President Trump's remarks about the white supremacist demonstration in Charlottesville ("good people on both sides") and Ken Frazier's resignation from the president's business advisory council in response thereto, see, *"At What Price a Seat at the Table,"* by Steven H. Holtzman and Jeremy Levin, August 16, 2017, https://leadershipandbiotechnology.blogspot.com/2017/08/at-what-price-seat-at-table.html.

2. For a further discussion, occasioned by the results of the 2016 election and the industry's preliminary political response thereto, see, *Leadership* and *Responsibility,* by Steven H. Holtzman, BioCentury, January 2, 2017, (pp.22–24). (Also available at: https://leadershipandbiotechnology.blogspot.com/2018/04/finally-posting-where-this-blog-started.html)

world. In brief, both arose in response to, and as a rebuke of, a world order in which what constituted truth and justice was the dictate of the powerful—whether that power was political, military, economic, or ecclesiastical. If you are doing science, you cannot exclude data because they are contrary to the prevailing dogma of the powerful; you cannot exclude the voice of another scientist because of their color or gender or nationality. Our industry's new normal must feature political activism in favor of those who support policies grounded in science, not ideology, and in social justice. We must equally raise our voices in opposition to those who would devalue the roles of science, data, and facts in public policy formation and who exclude participation based on racist, sexist, or other supremacist attitudes. Truth and justice, science and humanism: in each case, two sides of the same coin.

Commitment to human health

Our industry is devoted to developing new medicines to ease human suffering and engender human well-being. To use an example in which I was involved, when Biogen developed a novel, longer-lasting Factor VIII to treat hemophilia, we did not say to ourselves, "This will be great for the hemophiliac children of wealthy, white Americans of Northern European descent." Our commitment is to all of humanity, and our goals are fully achieved only when all who may benefit from the products of our labor do so. Moreover, while enabling people to have (relatively) pain- and disease-free lives is a good in and of itself, we also care about this because, in its absence, humans are not fully able to develop their potential and capabilities, to fully exercise their freedom and flourish. Our quest is incomplete unless and until there is universal accessibility to the medicines we develop, and all, when they have achieved health, have a chance to flourish. Our active political engagement should reflect this commitment.

Industry origins and founders

The biotechnology industry was born in the late-1970s. It decidedly was not a child of corporate America or the brainchild of corporate managers. The new industry's founders were academic scientists and entrepreneurs. They saw the potential of the recombinant DNA revolution to improve human health. This became their mission, their passion. And, as the 1970s became the 1980s, and the biotechnology industry grew, that mission attracted a next generation of scientific and business leaders who shared with the industry's founders that commitment, that goal, and a fundamental set of moral values.[3] Also noteworthy: unlike the leadership ranks in corporate America at that time, many of biotechnology's business and scientific leaders were immigrants or first-generation Americans.

Elsewhere[4], I have reflected on David Brooks's commentary on the distinction between having a career (something you choose) and having a vocation (something you are called to and cannot *not* do) and what this means for how leaders lead. I have suggested that to build a career, it is of singular importance not to make (or be seen to have made) a mistake; that fulfilling one's vocation, on the other hand, requires putting one's self at risk of being (publicly) wrong and taking responsibility for your error; that our industry's founders—on the whole—were driven by such a sense of vocation; and, that in critical moments when the most is at stake, the soul of a biotechnology leader is revealed as being driven either by risk aversion and personal gain or the industry's founding values and mission.

---

3. This stands in remarkable contrast to what was going on more broadly in corporate America in the 1980's. That decade marked the beginning of what has now been a forty-year period in which income and wealth inequality has radically increased, social programs for the less well-off have been under constant assault, taxation has become increasingly less progressive, and wealth has accumulated (as Thomas Pinketty points out in *Capital in the 21st Century*, to an historically unprecedented level) in the hands of those who manage money, as opposed to those who create primary value. This was, after all, the decade that brought us a fascinating collection of heroes and tales, factual and alternative-factual, e.g. Ivan Boesky, (the young) Donald Trump, Michael Milken, Gordon Gecko, *Bonfire of the Vanities*, and *Barbarians at the Gates*.

4. See, *On NPV, DCF and IRR(elevance)*, Part 3: *Moral of the Story*, Holtzman, BioCentury, September 26, 2016. (Also available at: https://leadershipandbiotechnology.blogspot.com/2018/08/on-npv-dcf-and-irrelevance.html).

And, what if we do not seize this opportunity to create a new normal for ourselves, do not seek to realign our actions with the better angels of *our* nature?

In 1846, Henry David Thoreau spent an evening in jail in protest of slavery, for refusing to pay the poll tax. As legend has it, Ralph Waldo Emerson visited Thoreau that evening and had the following exchange: "Henry, why are you here?" to which Thoreau replied, "Why are you *not* here?"

Thoreau was entreating Emerson to realize that who Emerson claimed to be in virtue of his Abolitionist principles and beliefs about the repugnance of slavery was belied by his own actions. Thus, Thoreau was not criticizing Emerson from Thoreau's perspective. Rather, he was inviting Emerson to see that he was failing *his own* better angels. Emerson was living in contradiction, and living in contradiction with one's own nature is not a state of factual or (merely) political error: it is the stuff of tragedy.

Leaders of our industry have the opportunity right now to shape the future of biotechnology in the time *after* COVID-19. That future can be one in which we act, and act publicly and assertively, in a manner consonant with *our* nature. The alternative, I fear, is not a return to a mere lack of public support, it is tragedy.

From 1985 through 2020, **Steven Holtzman** served as a co-founder and EVP of DNX Corporation, an early leader and CBO of Millennium Pharmaceuticals, a co-founder, CEO, and Chair of Infinity Pharmaceuticals, the EVP of Corporate Development of Biogen, and the first CEO of Decibel Therapeutics. He currently is a board member of Molecular Partners, the chair of the boards of Camp4 Therapeutics and Qihan Bio, and a board member of the Berklee College of Music.